Introducing Technology Computer-Aided Design (TCAD)

Introducing Technology Computer–Aided Design (TCAD)

Fundamentals, Simulations, and Applications

Chinmay K. Maiti

PAN STANFORD PUBLISHING

Published by

Pan Stanford Publishing Pte. Ltd.
Penthouse Level, Suntec Tower 3
8 Temasek Boulevard
Singapore 038988

Email: editorial@panstanford.com
Web: www.panstanford.com

British Library Cataloguing-in-Publication Data
A catalogue record for this book is available from the British Library.

ISBN 978-981-4745-51-2 (Hardcover)
ISBN 978-1-315-36450-6 (eBook)

Printed in Canada

Dedicated to

Late Dr. Rakhal Chandra Maiti

and

Dr. Kali Kinkar Das

Contents

Preface

At the beginning of 2015, the fifth-generation Intel Core processor was released, which is built with 14 nm technology containing 1.3 billion transistors. The first processor built with the 14 nm technology is the Intel Core-M processor. Generally speaking, the architecture of transistors has now switched from the planar to the vertical design. To provide the necessary support to keep the semiconductor industry on its technological growth track, heavy emphasis is being placed on technology computer-aided design (TCAD) now for achieving key long-term research results.

Even though the predominant focus of the semiconductor industry in the 1990s and early 2000 was on biaxially strained devices, the current focus has shifted to process-induced uniaxial stress, which is being adopted in all high-performance logic technologies. Uniaxial stress has several advantages over biaxial stress, such as larger mobility and performance enhancements. Encouraged by the strain-enhanced planar metal–oxide–semiconductor field-effect transistors (MOSFETs), researchers recently applied uniaxial stress to multigate devices with a metal gate and high-k dielectric as performance boosters. Despite the advantages provided by the manufacture of transistors, introduction and optimization of the mechanical stresses in the channel remain essential.

TCAD simulations provide a comprehensive way to capture the electrical behavior of different devices with different materials and structures for performance assessment. This monograph aims to provide the reader with a comprehensive understanding of the technology development for stress- and strain-engineered device processing. With specific case studies, applications of the process/device simulation programs in process and device development are shown. The aim of this monograph is also to provide device/circuit designers with the ability to predict what impact statistical process variations will have on their designs, as early as possible, using several commercially available tools. As currently, uniaxial strain is being adopted in all high-performance logic technologies and applications by the industry, we shall focus mainly on process-induced uniaxial stress in this monograph.

Next-generation wireless systems, driven by a vast assortment of rapidly emerging applications operating at radio frequency (RF) and millimeter-wave frequencies, are placing increasingly stringent cost and performance demands upon the supporting microelectronics technologies. On the basis of its performance capabilities, low cost, and capacity for high integration, silicon germanium (SiGe) heterojunction bipolar complementary metal-oxide semiconductor (BiCMOS) technology has established itself as a strong technology contender for a host of such circuit applications, including analog and mixed signals, RF, and millimeter waves. However, as operating frequencies for wireless applications are pushed upward in the spectrum, SiGe heterojunction bipolar transistor (HBT) technologies face significant challenges at the transistor level as operating voltage decreases and performance requirements increase.

Starting with the 90 nm technology node, improved transistor performance has been achieved with the introduction of stress and strain in the transistor. One of the major limitations of the currently available related books is that the most of the design and simulation results presented are obtained using 2D simulations without involving stress. Due to the ultrasmall size of state-of-the-art devices, 3D effects have been dominant. To achieve a better understanding of simulated and fabricated device characteristics, 3D process/device simulation involving stress is necessary. This monograph presents mostly the 3D simulation and analysis of stress- and strain-engineered semiconductor devices for digital, analog/RF, and power applications. Detailed and extensive TCAD simulations are carried out for 3D fin field effect transistors (finFETs), and the key parameters are identified.

TCAD is shown to be an excellent resource for teaching microelectronics. The objective of a laboratory component of any semiconductor fabrication course is to teach the students the unit processes involved in microelectronic fabrication and to introduce the practice of process development. Virtual wafer fabrication (VWF) has become an integral part of the semiconductor industry now. The possibility of teaching semiconductor manufacturing in a university environment in a highly cost-effective manner by taking the advantages of high-speed internet and available TCAD tools has been explored.

Recently, several excellent books and monographs have appeared on multigate MOSFETs, high-mobility substrates and Ge microelectronics, and strained semiconductor physics. Numerous papers have appeared on strained Si and process-induced strain, but there is a lack of a single text that combines both strain- and stress-engineered devices and their design and modeling using TCAD. Attempts have been made to summarize some of the latest efforts to reveal the advantages that strain and stress have brought in the development of strain-engineered devices. We have included important works by the research community, as well as our own research students' works and ideas. The monograph is mainly intended for final-year undergraduate and postgraduate students of electrical and electronic engineering disciplines and scientists and engineers involved in research and development of high-performance devices and circuits. The monograph may serve as a reference on strain-engineered heterostructure devices for engineers involved in advanced device and process design. Instructors involved in teaching microelectronics may also find this monograph useful for laboratory education using remote web-based TCAD laboratories.

After reading the monograph, the students will learn the fundamentals of process and device simulation programs. They will be able to analyze arbitrary device structures to speed up the designing using commercial software. Approaches presented in this monograph are expected also to boost the use of TCAD tools for device characterization and compact model generation.

I am extremely grateful to my research students who made significant contributions to make this monograph a reality. I would also like to express my deep appreciation for the Pan Stanford Publishing team. Finally, I would like to thank my family members (wife, Bhaswati, and sons, Ananda and Anindya) for their support, patience, and understanding during the preparation of the manuscript.

C K Maiti
SOA University
Bhubaneswar
September 2016

Chapter 1

Introduction

For the last five decades, CMOS and bipolar technologies have provided consistent scaling and enabled the implementation of high-density, high-speed, and low-power ULSI systems. During the 1960s and until early 2000, continuous enhancement of MOSFET performance has been achieved mostly by geometrical scaling, as was predicted by Gordon Moore in 1965. Device performance enhancement has been achieved by combining gate oxide thickness and gate length scaling. The microelectronics industry has relied mainly on the shrinking of transistor geometries for improvements in circuit performance and cost per function over the decades. Since the 2000s, MOSFET performance enhancements have been achieved mainly via device (stress and strain) engineering, while keeping the gate dielectric thickness almost constant.

In the semiconductor industry, each new technology generation is driving toward higher performance and increased complexity, which gives rise to higher costs of product development. Over the last 60 years, device scaling has been relatively straightforward using geometrical scaling. Further downscaling of devices has approached the physical limit and the technology cycle has become slow. Figure 1.1 illustrates the trend of cost per device, transistor leakage, on-state current, and transistor switching speed as the gate length, gate width, and gate dielectric thickness shrink. The aggressive scaling of transistors has reduced equivalent oxide thickness of gates to less

Introducing Technology Computer-Aided Design (TCAD): Fundamentals, Simulations, and Applications
C. K. Maiti
Copyright © 2017 Pan Stanford Publishing Pte. Ltd.
ISBN 978-981-4745-51-2 (Hardcover), 978-1-315-36450-6 (eBook)
www.panstanford.com

than 1 nm, and it is expected to be 0.4 nm by 2028. The gate length is also expected to be simultaneously scaled down from 20 nm to 5 nm.

Further improved performance of integrated circuits (ICs) may be obtained by fundamental process improvement of the transistors in them—via new device structure and doping strategies and shrinking (or downscaling) the transistors' size, which increases their speed and reduces power consumption. The semiconductor industry has been looking for alternate performance boosters, in particular, by introducing new materials and new device architectures in place of traditional standard silicon CMOS and bipolar technology. A new era of CMOS technology has begun with the introduction of 22 nm bulk fin-shaped field-effect transistors (finFETs) by Intel, continuously driving the future CMOS scaling toward low power.

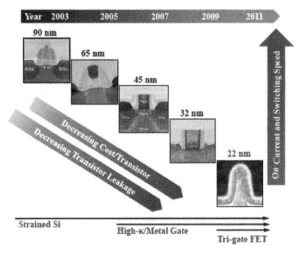

Figure 1.1 Technology trend of CMOS scaling in terms of device dimension and performance. Cross-sectional TEM images of transistors for technology nodes from 90 nm to 22 nm are shown here [11].

With each generation of scaled technology the transistor performance has improved. The transistor architecture has undergone significant changes from conventional MOSFETs with the introduction of strained silicon at the 90 nm node [1] and a HfO_2 gate dielectric (replacing SiO_2) with a metal gate (replacing polysilicon) at the 45 nm technology node. Starting with the 90 nm technology node, improved transistor performance has been achieved with the introduction of stress into the transistor channel region. Use of intentional and unintentional stress/strain to boost device performance and reduce size and power has been very important in semiconductor manufacturing. To continue with the Moore's law trend, incorporation of technology boosters such as strained Si, high-*k* gate dielectric, and metal gate have been introduced. It is believed that the semiconductor device will continue downscaling to the 7 nm gate length by the year 2018. Most of the emerging transistor device technologies, such as finFETs and multigate transistors are based on SOI material.

Strain technology, which employs mechanical stress to alter the band structure of silicon and reduces the carrier effective mass and scattering rate, is introduced to increase carrier mobility to continue scaling. As stress has a major impact on transistor characteristics in modern processes, stress effects need to be considered in the simulation in order to study the influence of process variations from stress-related effects. Figure 1.2 shows the various stress/strain techniques currently in use in the microelectronics industry.

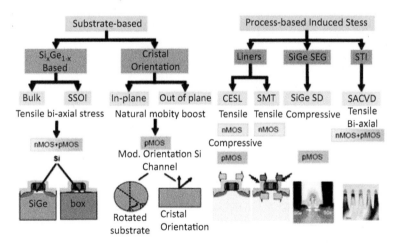

Figure 1.2   Schemes of front-end stressor types currently in use in the industry [12].

The application of new materials and/or technology requires considerable time and cost because there are so many options and combinations for the development. In this respect, exploiting technology computer-aided design (TCAD) has become more important for the virtual characterization of technology in advance. Recent research efforts have helped us to better understand the issues specific to transistor design, and we are currently better equipped with process technologies. Also, with the perceived end of conventional scaling on the horizon, drastic changes in the device structure and/or materials may be the only way to continue with the past trends. Obviously, any new transistor design should be thoroughly evaluated and benchmarked to existing state-of-the-art technology. Figure 1.3 shows the envisaged future for devices and systems. In contrast to the past design rules, now systems are more and more integrated in a single part. To achieve such integration, novel processes and technologies need to be conceived.

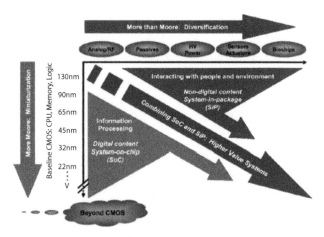

Figure 1.3 "More than Moore" principle (International Technology Roadmap for Semiconductors, http://www.itrs.net/).

TCAD may be used for computer-aided design of semiconductor devices, fabrication process design and development, technology characterization for circuit design, manufacturing yield optimization and process centering, and computer-integrated manufacturing. Complementing experimental work, the TCAD combination allows optimizing and predicting the performance of new devices. It is therefore appropriate to use TCAD tools to study comprehensively the necessary requirements for realizing the ultimate limit of CMOS and bipolar scaling, which poses serious technology and design challenges. Simulation is now an essential tool for technology development. In fact, numerical modeling is the only way to provide comprehensive investigations of the front-end-of-line (FEOL) process-induced strain in silicon using piezoresistive-related physics.

In this chapter, an overview of the evolution, role, need, and advantages of TCAD tools as an aid toward process design, device simulation, and fabrication of modern ICs, device characterization, and compact model generation is given. Since 1990, a major effort organized by the TCAD subcommittee of the CAD Framework Initiative has been launched to develop an industry standard for TCAD. Excellent overviews on various TCAD framework activities can be found in Refs. [2, 3].

1.1 The Need

According to the International Technology Roadmap for Semiconductors (ITRS), technology computer-aided design (TCAD) can lower technology development costs up to 40% by reducing the number of experimental lots and shortening development time. This is significant, considering the rising costs of product development and new wafer fabrication facilities. It is important to develop an improved understanding of the correlation between process variables and electrical device parameters. TCAD tools provide insight into many of the physical effects that impact manufacturability and yield. Through proper calibration with prototype wafers, TCAD can accurately predict the behavior of devices for a new technology node.

TCAD-based transistor design can be used at the early stage of technology development to introduce accurate statistical variability and reliability information in the process design kit (PDK). This information is of great importance for designers and can reduce the time-to-market, avoid overdesign, increase yield, and reduce the expensive chip design process costs. The narrowing of the gap between front-end and backend manufacturing processes is a typical example of the new challenges in the electronics market and design tools [4, 5].

In recent years, the coverage of the interaction between design and process engineers has been extended to various aspects with the evolution of design for manufacturing (DFM) and design for reliability (DFR) methodologies. In combination with wafer data, the strength of TCAD lies in the prediction of device and interconnects variability due to layout as well as random variations during fabrication. The knowledge gained through TCAD can be encapsulated into appropriate models and efficiently used for process optimization, ultimately leading to more robust designs. Variability information can also be incorporated into design tools through statistical compact models. Ultimately, this leads to an improved design flow that addresses parametric yield and manufacturability issues in a comprehensive way.

1.2 Role of TCAD

To understand the challenges facing advanced device design, one must first understand the design parameters available to a device designer and their significance. The goal of device design is to obtain devices with high performance, low power consumption, low cost, and high reliability. Figure 1.4 shows the traditional device design flow. Instead of going through an expensive and time-consuming fabrication process, computer simulations can be used to predict the electrical characteristics of a device design quickly and cheaply. TCAD can also be used for reducing design costs, improving device design productivity, and obtaining better device and technology designs (Fig. 1.5). TCAD consists mainly of two parts, process design and device simulation.

1.3 TCAD: Challenges

TCAD is now an indispensable part of semiconductor modeling and design. In the semiconductor industry, the core production process of manufacturing integrated circuits (ICs) takes place in a front-end manufacturing facility. The wafers go through a specific processing sequence that consists of a certain number of processing steps. Modern complex semiconductor manufacturing processes consist of a large number of processing steps, long processing time, dynamic interactions among different tools, and complex interrelations between tool performances and product qualities. Semiconductor manufacturing usually involves hundreds of processing steps, months of processing time, reentrant process flows, and unpredictable relationships between tool performances and yield. As semiconductor processes become more complex (e.g., for a fin field-effect transistor [finFET] structure) and computational algorithms more sophisticated, usability becomes a more severe problem. Also, accurate TCAD simulations and modeling of physical devices depend critically on calibrated physical models and proper input data. With the scaling down of complementary metal-oxide semiconductor (CMOS) and bipolar technology, many secondary effects become more pronounced, such as drain-induced barrier lowering, process variations, gate tunneling leakage, mobility degradation, and increasing power consumption. These secondary effects lead to dramatic challenges in robust circuit design and system integration.

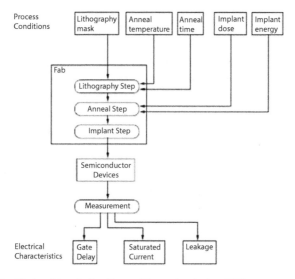

Figure 1.4   Technology design flow without simulation [13].

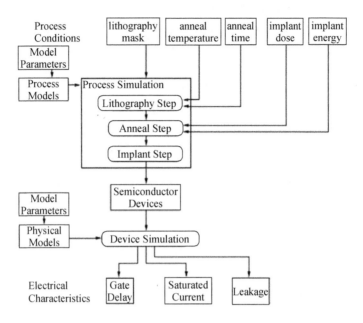

Figure 1.5 TCAD-based technology design flow [13].

As the device densities are growing exponentially, fundamental physical limits are being approached and design constraints are becoming serious issues. The conventional way of describing, designing, modeling, and simulating such nanoscale semiconductor devices and the assumption of continuous ionized dopant charge and smooth boundaries and interfaces is no longer valid. As a consequence, TCAD is being challenged by a number of fundamental problems such as microscopic diffusion mechanisms, quantum mechanical transport, molecular dynamics, quantum chemistry, and high-frequency interconnect simulation. Today's metal–oxide–semiconductor field-effect transistors (MOSFETs) are truly atomistic as they have reached decananometer (between 10 and 100 nm) dimensions with 40–50 nm physical gate length devices in the 90 nm technology node, and 10 nm MOSFETs have been demonstrated in a research environment. It is expected that in 2016, the physical dimensions of MOSFETs in the production environment will reach 9 nm. Some of the main challenges for TCAD to successfully deal with current process/device simulation problems are:

- Stress/strain can influence point defect concentrations and migration, dopant activation, and defect nucleation and evolution. All these issues and new materials need to be included in process simulation to maintain its usefulness for future technology nodes. In further scaling, stress sources will be in closer proximity to each other and the silicon body will likely be thinner.

- Modern manufacturing systems require adaptability to rapidly changing environments. In particular, frequent changes in the products, processes, or equipment in a semiconductor processing facility reduce the value of experiential models used in more stable environments.

1.4   TCAD: 2D versus 3D

Aggressive scaling results in complex physical phenomena in contributing to the device behavior, and small dimensions severely limit the descriptive capabilities of measurements. Device downsizing has also led to the introduction of new materials and architectures, thus complicating the transistor structure. TCAD as a complement to experimental runs provides a more comprehensive way to characterize technologies and to optimize their performance. The application space of current TCAD tools includes the development of semiconductor technologies—from deep-submicron logic, memory, solar cells, and sensors to compound semiconductors and optoelectronics. More and more complex device modeling is needed for computer simulations at the process and physical levels. Considering device processing difficulties, device design and optimization need to be performed through device physics analysis.

3D integration technology is highly compatible with existing CMOS manufacturing processes, thereby making 3D technology a favorable candidate in the IC industry. 3D integration technology offers several advantages to increase performance and functionality, while reducing cost. In 3D technologies, multiple dies are stacked in a monolithic fashion where the communication among the dies is achieved by vertical through-silicon vias (TSVs). Despite important advantages, however, 3D integration has certain challenges that need to be addressed. The traditional role of TCAD has been the continuum-based large-scale 2D/3D simulation. To achieve a better matching between simulated and fabricated device characteristics, 3D process/device simulators need to be employed as the verification tool to validate simulation results.

1.5   TCAD: Design Flow

The shortening of the design-to-manufacture time for a new product mandates a rapid design cycle for bringing up new technologies. The proliferation of TCAD tools has thus a significant economic impact on product development. In fact, an effective use of TCAD tools saves much experimental time for calibrating process and device parameters and minimizes the number of trial-and-error iterations. To develop a general framework and a systematic approach to the use of TCAD in aiding technology development and circuit design, it is important to understand the nature of TCAD and to make the best use of it. With the maturity of TCAD tools, real wafer fabrication (RWF) can be emulated by process simulation, from which realistic device structures and doping profiles can be generated, and transistor performance can be characterized through device simulation with reasonable accuracy. The multilevel design flow for fabricating ICs is shown in Fig. 1.6.

Simulation Program with Integrated Circuit Emphasis (SPICE) parameters can be extracted from the virtual device I–V characteristics for circuit simulators and timing analyzers. Interconnect delays can also be extracted through technology characterization, which can provide information for design rule checker (DRC) and layout parasitic extraction (LPE) tools in the physical design. Advanced simulation capability greatly reduces the cost by (i) detecting design flaws in the early design stage and (ii) achieving optimal solutions through parameter calibration.

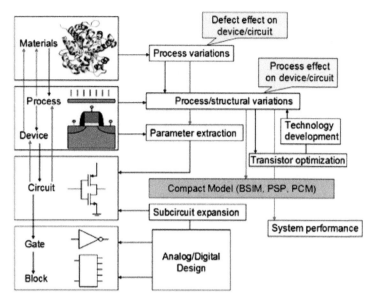

Figure 1.6 Compact multilevel technology/device/subsystem modeling flow. After Ref. [14].

1.6 Extending TCAD

Current TCAD use is mostly limited to process and device simulations; however, it may be extended for the development of compact models suitable for circuit and system-level analysis. The increasingly competitive IC market requires new design approaches using TCAD. Fabless companies also need to quantify the final product sensitivity to process parameters and to tailor them for specific application, aiming at performance optimization for large-volume production where TCAD can play an important role. From this point of view, the traditional role of TCAD needs to be extended to the advanced high-level approach and the fundamental low-level approach as well. TCAD needs to be extended also for studying process variability and using the process parameter analysis in the very large-scale integration (VLSI) circuit performance evaluation.

Circuit simulation is indispensable for the design of ICs, and a compact device model is an essential part of the circuit simulation. Technology scaling is increasing the complexity and nonideality of the electrical behavior of semiconductor devices. Besides providing a deep insight, especially for aggressively scaled devices, TCAD simulations may help in the generation of predictive models that play a crucial role in reducing circuit development time and cost for the semiconductor industry. Device modeling also plays an important role in semiconductor fabrication, especially for the growing diversity of device requirements.

Compact models generally include SPICE-like parameters obtained from the device electrical behavior. Also, variations of SPICE-like parameters need to be carefully estimated to achieve acceptable model predictivity, including process yield evaluation. As such, compact model extraction tools are becoming very important. In the following section, we introduce several such tools available in the public domain. Mystic [6] is a compact model extraction tool specifically designed for accurate statistical compact model extraction. It provides a scripted environment that enables the development of advanced devices based on data from TCAD simulations or silicon measurement. Mystic is applicable to bulk CMOS, finFET, and fully depleted SOI processes.

DFR is becoming indispensable for every aspect of product quality assurance, and it needs be implemented in the design flow via various levels of abstraction. As a low-level approach, it is important to understand the fundamental physics involved and to incorporate them into models. There is a growing need to understand the effect of process variation on circuit performance. Using calibrated TCAD simulations it is possible to study the effects of slight process variation on device characteristics. Process-aware SPICE models offer a way to bring process variation information to the design sphere. Using TCAD tools, many reliability issues can also be studied quantitatively. Examples are hot-carrier degradation of interfaces, threshold voltage shifts during negative-bias temperature instability (NBTI) stress, radiation effects and soft errors, electrostatic discharge (ESD) and latch-up, thermomechanical issues, electromigration, and stress voiding.

1.7 Process Compact Model

Design complexity is getting constantly enhanced due to the increasing density of integration (more than a billion components), which has led to a huge gap between physical simulation at the nanoscale and earlier large-geometry IC design. Along with this come significant parametric variations in the devices due to very small feature sizes. TCAD for the manufacturing paradigm effectively strengthens the connection between technology development and manufacturing. TCAD modeling algorithms need to take care of the interaction between design layout and manufacturing processes, which creates layout pattern–dependent systematic yield models that encompass process technologies such as lithography, etch, and interlayer dielectric chemical-mechanical polishing (CMP), copper CMP, and shallow trench isolation CMP. Yield optimization has become the single-most important factor for reducing the product cost. Using a process compact model (PCM), manufacturing engineers can analyze process sensitivity, recenter a process, and identify key process steps to improve overall process capability to improve the yield. The TCAD-for-manufacturing (TFM) paradigm effectively strengthens the connection between technology development and manufacturing.

1.8 Process-Aware Design

A systematic methodology also needs to be developed for identifying and characterizing the main sources of transistor performance variation. Paramos [7] from Synopsys is one of the first automated software programs that links process and circuit simulation. It stands as an example of what should be the new bridge between design and manufacturing, a process-aware transistor model. Process-aware device modeling involves the study of the impact of geometrical variation (e.g., nonuniform lateral dimension, nonideal cross section, atomic-level dopant fluctuation) on the $I–V$ characteristics, including quantum effects. Techniques for evaluating variability at an affordable computational cost are being sought, in addition to increasing complexity of the device structure and behavior. One needs to develop models based on an orthogonal set of parameters that correlate the geometry data to electrical characteristics.

1.9 Design for Manufacturing

DFM consists of three stages: physical design, resolution enhancement techniques, and design-driven techniques. Predictive potentials of TCAD depend on process variations that get increasingly critical with device downscaling into the nanometer range. For example, phenomena such as line edge roughness (LER) and random dopant fluctuations (RDFs) broaden the device parameter distributions, thus requiring statistical analysis. DFM and design for yield (DFY) require EDA tools that fully comprehend the impact of novel technology concepts and their influence on the process variability of devices and interconnects.

A collaborative platform for DFM aims to meet this challenge by joining process and circuit simulators as well as a set of process characterization experiments that are needed to enable quantitative DFM. At the 22 nm node and beyond, process variability will increase with feature scaling and the introduction of new materials and techniques such as strain engineering. TCAD is now being given a major emphasis in DFM. TCAD addresses variability as it complements silicon metrology data with accurate, calibrated process and device models. This is inherently a multidisciplinary problem that requires a process/device/design framework for an optimal solution.

1.10 TCAD Calibration

Considering the diverse requirements, it is natural to ask for the level of confidence at which process simulation is useful for process development. In fact, there are generally delays between the technological concepts, the qualitative understanding of their effects, and finally the quantitative reproduction within TCAD. For accurate device characterization the process and device simulators in the TCAD suite of tools must be calibrated.

Written from an engineering application standpoint, this monograph provides the background and physical insight needed to understand new and future developments in the modeling and design of devices at the nanoscale. The focus of this monograph is on the state-of-the-art devices, implemented in high-mobility substrates such as GaAs, GaN, silicon-germanium (SiGe), strained Si, Ge, and ultrathin germanium-on-insulator platforms, combined with high-k insulators and metal gates. We offer plenty of 3D examples with step-by-step procedures, making the monograph a useful reference for those who wish to set up their 3D simulation using any TCAD tool with 3D capability. Important features discussed in modeling and simulations provide a survey of the challenging issues for TCAD of the next decade with an emphasis on the fundamentals of the models themselves. Several examples with different levels of difficulty are provided for illustration of the basic concepts, while others deal with the state-of-the art technology illustrations. The monograph provides a balance between the basic conventional technology concepts, equations, physics, and recently developed technologies through TCAD simulation. Most of the simulation results presented throughout this monograph were performed using Silvaco and Synopsys TCAD tools.

The monograph consists of 14 chapters (substrate-induced strain engineering in CMOS to bipolar technology, process-induced stress, electronic properties of strained semiconductors, stress-engineered MOSFETs, noise in SiGe heterojunction bipolar transistors (HBTs), strain-engineered HBTs, process compact modeling, and process-aware design of strain-engineered devices) and looks ahead of the 22 nm node.

1.11   TCAD Tools

The main objective of this chapter is to introduce the advantages of TCAD simulations for device and process technology development and characterization, to introduce the fundamental physics and mathematics involved with TCAD tools, and to expose readers to the most popular commercial TCAD simulation tools such as from Silvaco and Synopsys. Various tools used in this monograph for simulation purposes are introduced in a comprehensive manner. As the mainstream CMOS technology is scaled below the 22 nm technology node, development of a rigorous physical and predictive compact model for circuit simulation that covers geometry, bias, temperature, DC, AC, radio frequency (RF), and noise characteristics becomes a major challenge. In this chapter we also discuss the fundamentals of the mathematical model of elasticity, define a system of governing equations and corresponding boundary conditions, and describe sources of strain/stress and material parameters used for stress calculations in isotropic and anisotropic materials.

1.12   Technology Boosters

Even though the predominant focus of the industry in the 1980s and 1990s was on biaxially strained silicon devices [8, 9], the current focus has shifted to uniaxial stress. Uniaxial stress has several advantages over biaxial stress, such as larger mobility enhancements and smaller shifts in threshold voltage [10]. Encouraged by strain-enhanced planar MOSFETs, researchers recently applied uniaxial stress to multigate devices with a metal gate and high-k dielectric as a performance booster. With each generation of technology node, the transistor performance has been improved. But, it's only from the 90 nm technology node that some notable changes have been made to the traditional transistor architecture, like strained Si and a high-k/metal gate at the 45 nm node.

The current MOSFET technology is moving toward strained devices to optimize mobility. This requires the use of alloys. From a process modeling point of view, SiGe materials offer several new challenges to our understanding. Most practical device structures use the lattice mismatch between SiGe and Si to create strain. This means relevant devices are mingling two things, strain and chemical composition, both of which can alter activation, diffusion, and defects. Stress effects will become more important with device scaling and with novel structures. In modern processes, stress is difficult to characterize due to nonavailability of nanometer-scale probing to measure mechanical stress. While introducing new devices structures, innovation has always been an important part in device scaling and the integration of new materials. It is envisioned that the right combination of global biaxial and local uniaxial strain could provide additional mobility improvements.

1.13   BiCMOS Process Simulation

Despite the overall industry domination of CMOS logic, there remain high-performance niche markets where other transistor architectures can still thrive. This is particularly true for the realm of analog electronics. CMOS and bipolar technologies have their weak and strong points. CMOS has qualified to be the most appropriate choice for VLSI applications because of its low DC power dissipation and its high packing density, yet its speed is limited by the capacitive loading. On the other hand, bipolar digital circuits outperform CMOS in terms of speed but are power consuming. The existence of this gap implies that neither CMOS nor ECL bipolar have the flexibility required to cover the full delay power space. This can only be achieved by a technology such as bipolar complementary metal-oxide semiconductor (BiCMOS). In this chapter, we introduce simple CMOS and bipolar process and device simulation. Fully 3D modeling approaches for MOSFETs are also demonstrated to represent the complicated structure and behavior of aggressively scaled devices.

1.14   SiGe and SiGeC HBTs

Beyond CMOS, bipolar-based technologies target ever-increasing analog and mixed-signal IC performance with a strong impact on parametric yield. Not surprisingly, there is no single technology that gives optimum performance in all aspects. Thus, for a specific application, one needs to choose from various available technologies. For example, for low-frequency applications up to 2.5 GHz, Si-based BiCMOS technology could be the best choice. However, for very-high-frequency applications, GaAs-based devices could be the choice due to their superior high-frequency and breakdown characteristics. With steady performance gains and continued innovation in process integration, SiGe HBT BiCMOS technology is becoming an increasingly viable and affordable solution for highly integrated, high-performance mixed-signal applications. In this chapter, process and device simulations of advanced SiGe and SiGeC HBTs are considered.

1.15   Silicon Hetero-FETs

The technology-scaling trend leads to increasing complexity of ICs. To boost MOSFET electrical performances, stress and strain engineering is nowadays used in all advanced semiconductor technologies. Most of semiconductor device manufacturers such as Intel, IBM, and TSMC are using mechanical stress to enhance the performance of nano-CMOS transistors at 45 nm and below. Stress is used to increase carrier mobility in the channel [10]. SiGe with a biaxial compressive strain has been demonstrated to be favorable for hole confinement and hole mobility enhancement because of its band offset and split in the valence band. SiGe is the ideal material to boost the speed of both n- and p-channel Si MOSFETs. In this chapter, we consider the design of strained SiGe and strained Si channel MOSFET designs.

1.16   FinFETs

Over the past decade, transistor structures have evolved a step further from planar, classical, single-gate FETs to 3D multigate FETs whose behavior can only be fully explained by advanced carrier transport phenomena. Planar CMOS technology has reached its scaling limits at the 22 nm node, where it is increasingly difficult to design high-performance, low-power devices with good yield in the presence of global and local process variations. Multigate FET technology is the best alternative that can extend scaling to the sub-10 nm technology nodes with minimum additional processing costs. FinFETs may replace conventional CMOS devices in the future technology generations due to their intrinsically better scalability. From the fabrication perspective, the most likely candidate for widespread adoption amongst the multigate devices is the finFET. In this chapter, TCAD techniques are used to study the design of finFETs.

1.17   Advanced Devices

For applications beyond the 22 nm node trigate/multigate devices, Ge and III–V MOSFETs, quantum well and tunnel FETs, and a host of other novel devices are being explored. Intuitive physical analyses yield valuable insights and allow easy and qualitative comparison of the different device structures. In this chapter, advanced devices such as ultrathin-body silicon-on-insulator (UTBSOI), high-electron-mobility transistors, AlGaN/GaN HFETs, and high-power SiC devices are considered.

1.18   Memory Devices

Nonvolatile semiconductor memory is an increasingly popular type of semiconductor memory because of its ability to retain data without external power. Because of its high-performance capabilities and applications in a wide range of popular consumer products, Flash memory is a very important member of the nonvolatile semiconductor memory family. Nonvolatile memory has drawn much attention over the past years due to their its applications in the consumer electronics market where memory devices with a retention time of ~10 years are desired. The main failure mechanism in Flash-based memory devices is the threshold voltage shift and memory window narrowing caused by oxide degradation (due to charge trapping caused by electronic stress during program/erase cycling) and is of great research interest. In this chapter, design and simulation of memory devices are presented.

1.19   Power Devices

Over the last decade, there has been a growing research interest in the area of power electronic applications. Conventional MOS structures are also not suitable, especially in medium- and high-voltage power ICs as both small gate length and thin oxide thickness are related to the breakdown. To alleviate the effect of the electrical field on the gate oxide, several power MOS device structures such as lateral double-diffused MOS transistors (LDMOS) and vertically diffused MOS (VDMOS) have been proposed to greatly increase the breakdown voltage. The physical effects, self-heating and impact ionization, which are essential for modeling power devices, can be modeled using the corresponding state-of-the-art models. Unstable problems like snap-back, secondary breakdown, and thermal runaway effects can be simulated. In this chapter, we shall present the results of 3D process/device simulation of LDMOS and VDMOS devices.

1.20   Solar Cells

The first homojunction silicon solar cell developed in 1954 had an efficiency of only 6%. But the progress in silicon technology resulted in single-crystal silicon solar cells with high reliability and efficiencies reaching above 25%. Thin-film solar cells have become a strong competitor for single-crystal and polycrystalline silicon solar cells because of the cheaper raw material as well as processing costs. Substrates used in thin-film solar cells are a metal or a glass or a polymer. These are relatively cheaper compared to silicon wafers. In this chapter, TCAD of organic solar cells and tandem solar cells will be considered.

1.21   TCAD for SPICE Parameter Extraction

Strain technology has been successfully integrated into CMOS fabrication to enhance carrier transport properties since the 90 nm node. However, since the stress is nonuniformly distributed in the channel, the enhancement in carrier mobility, velocity, and threshold voltage shift strongly depend on circuit layout patterns, leading to extra process variations. Thus, a compact stress model is needed that physically captures this behavior to bridge the process technology with design optimization. With its physical basis that is capable of capturing detailed process effects, calibrated TCAD flows can be used to generate computationally efficient PCMs that retain key process-to-device correlations. In this chapter, extension of TCAD to extract SPICE parameters of transistors is discussed.

1.22   TCAD for DFM

TCAD is a bridge between the design world and the manufacturing world. Process information needs to be communicated into the design phase in such a way that informed design trade-offs can be made consistently. An understanding of the predictive modeling principles to gain insights into future technology trends is important for future circuit design research and IC development. The concepts of process compact and process technology modeling are essential to achieve the necessary knowledge transfer that has proven to be useful in the manufacturing world. The ultimate goal of predictive technology and process compact modeling is to describe any process technology accurately. The compact models are useful not only for long-term product design but also for early evaluation of a technology for circuit manufacturing. In this chapter, PCM generation is demonstrated using the Synopsys PCM tool.

1.23   VWF and Online Laboratory

In most undergraduate electrical engineering and virtually all graduate programs, there are courses on device physics and processing technology based on standard textbooks on MOS and bipolar device physics and have a very strong connection to the device modeling area and typical process technology. Associated with the subject, for simulation, the students generally use device/process simulation tools and perform circuit simulation with probably no interactions between the process/technology and device parameters.

A new conceptual framework for teaching microelectronics technology and device design is needed. All aspects of technology cross sections can be illustrated by using TCAD in process/device analysis and design via the Internet. The remote, web-based experimentation also can augment the laboratory experience of the students by offering access to sophisticated instrumentation. User-friendly, computer-controlled instrumentation and data analysis techniques are revolutionizing the way measurements are now being made, allowing nearly instantaneous comparison between theoretical predictions, simulations, and actual experimental results. Once this integration happens, it will no longer be crucial to have a piece of equipment physically located next to an engineer or scientist, thus opening the door for remote access via the Internet.

How micro- and nanoelectronic subjects can be taught efficiently and cost-effectively by using process and device simulation tools without any physical processing laboratory setup is demonstrated in this chapter. An integrated, remote, hardware-based online laboratory system for extending micro- and nanoelectronics education via characterization of semiconductor devices and their SPICE parameter extraction is described. The developed laboratory system combines both the convenience of remote and the effectiveness of traditional physical campus laboratories. As an educational tool the laboratory platform will enable students who do not have access to hardware laboratories to complement their theoretical knowledge by carrying out experiments remotely, performing simulations and comparing simulations with measured data using equipment located anywhere in the world and at any time of the day.

1.24 Summary

In this chapter, we briefly described the role, benefits, capabilities, and future perspectives of TCAD applications for semiconductor technology and provided an overview of TCAD usage. The evolution of modern TCAD and its challenges are discussed. Next-generation TCAD will play a key role in quantifying potential roadblocks, indicating new solutions, and keeping technology development on the path of Moore's law. Use of TCAD is shown as a crucial enabling methodology in supporting semiconductor technology progress. TCAD applications for the semiconductor technology development and the role of TCAD as a design-technology interface have been discussed in detail. Some of the main challenges for TCAD to successfully deal with process/device simulation problems are highlighted.

The purpose of the monograph is to bring various aspects of TCAD for microelectronic applications into one resource, presenting a comprehensive perspective of the field. The aim is also to provide students with device and process simulation examples applied to a number of different technologies, illustrating a quantitative link between the basic technological parameters and electrical behavior of microelectronic devices. Finally, in Chapter 14, we introduce the reader to an advanced tool (remote, hardware-based, web-based online laboratory) now available for extending the microelectronics education.

References

1. Maiti, C. K., and Maiti, T. K. (2012). *Strain-Engineered MOSFETs* (CRC Press, Taylor and Francis, USA).

2. (1995). *Microelectron. J.*, special issue, **26**(2–3), 77–315.

3. (1990). *Solid-State Electron.*, special issue, **33**(6), 591–791.

4. Dutton, R. W., and Yu, Z. (1993). *Technology CAD Computer Simulation of IC Processes and Devices* (Kluwer Academic, USA).

5. Armstrong, G. A., and Maiti, C. K. (2008). *TCAD for Si, SiGe and GaAs Integrated Circuits* (Institution of Engineering and Technology [IET], UK).

6. Garand, Mystic, 2013, http://www.goldstandardsimulations.com/.

7. Synopsys Inc., *Paramos User Guide*, Version J-2015.03, March 2015.

8. Maiti, C. K., and Armstrong, G. A. (2001). *Applications of Silicon-Germanium Heterostructure Devices* (Institute of Physics Publishing [IOP], UK).

9. Maiti, C. K., Chakrabarti, N. B., and Ray, S. K. (2001). *Silicon Heterostructures: Materials and Devices* (Institute of Electrical Engineers [IEE], UK).

10. Maiti, C. K., Chattopadhyay, S., and Bera, L. K. (2007). *Strained-Si Heterostructure Field-Effect Devices* (CRC Press, Taylor and Francis, USA).

11. Ran, C. (2014). *Strain Engineering for Advanced Silicon, Germanium and Germanium-Tin Transistors*, PhD thesis, National University of Singapore.

12. Fiori, V. (2010). *How Do Mechanics and Thermomechanics Affect Microelectronic Products: Some Residual Stress and Strain Effects, Investigations and Industrial Management*, PhD thesis, L'institut National des Sciences Appliquées de Lyon.

13. Kwong, M. Y. (2002). *Impact of Extension Lateral Doping Abruptness on Deep Submicron Device Performance*, PhD thesis, Stanford University.

14. Maiti, T. K. (2009). *Process-Induced Stress Engineering in Silicon CMOS Technology*, PhD thesis, Jadavpur University.

Chapter 2

Technology CAD Tools

In technology computer-aided design (TCAD), a broad range of modeling and analysis activities that consist of detailed simulation of IC fabrication processes, device electrical performance, and extraction of device parameters for equivalent circuit models are involved. TCAD tools provide detailed physical insight to achieve the optimum process performance affecting yield. TCAD represents our physical understanding of processes and devices in terms of computer models of semiconductor physics. The conventional role of TCAD in IC processing is shown in Fig. 2.1.

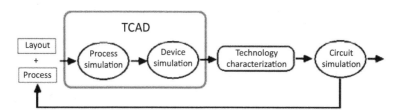

Figure 2.1 Role of conventional technology CAD in IC processing—the simple picture.

While advances in physical modeling of semiconductor devices and numerical methods have resulted in more accurate and robust TCAD tools, the usability, maintainability, and extensibility of these tools still remain a problem. The following is a brief introduction to

Introducing Technology Computer-Aided Design (TCAD): Fundamentals, Simulations, and Applications
C. K. Maiti
Copyright © 2017 Pan Stanford Publishing Pte. Ltd.
ISBN 978-981-4745-51-2 (Hardcover), 978-1-315-36450-6 (eBook)
www.panstanford.com

the evolution of technology and TCAD simulations, where the role of computer-aided design (CAD) and the increasing dimension of simulation problems are highlighted. Next-generation TCAD can and will play a key role in quantifying potential roadblocks, indicating new solutions, and keeping technology development on the path of Moore's law. This chapter offers insight into the physical basis of TCAD to allow users to leverage these tools effectively. We provide a detailed survey of the challenging issues for TCAD, with an emphasis on the fundamentals of the models themselves. Figure 2.2 shows the typical steps involved in computational device research and development with TCAD.

The scope of TCAD includes:

- Front-end process modeling and simulation, such as implant, diffusion, oxidation, etc.
- Device modeling and simulation for *I–V*, *C–V*, etc., characteristic generation
- Topography modeling and simulation
- Device modeling for circuit simulation
- Compact Simulation Program with Integrated Circuit Emphasis (SPICE) modeling

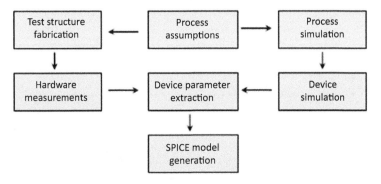

Figure 2.2   TCAD to SPICE model generation flow [27].

2.1   History of Process and Device Simulation Tools

In the 1970s, 1D approaches were generally sufficient to deal with bipolar technology and early metal-oxide-semiconductor field-effect transistors (MOSFETs). Use of 1D charge transport phenomena was predominant in large-geometry devices. Extrapolation of quasi-2D doping distributions from sets of 1D profiles helped process design optimization. Since the 1980s, aggressive MOS scaling led to the very-large- and ultralarge-scale integration (VLSI and ULSI) era based on complementary metal-oxide-semiconductor (CMOS) technology. Fully 2D simulators soon became indispensable to model, increasing process complexity and coupled physical effects, including local oxide isolation (local oxidation of silicon [LOCOS]), dopant diffusion, subthreshold conduction, and parasitic phenomena such as latch-up and punch-through. Technology computer-aided design (TCAD) products should enable users to explore and optimize easily the broad spectrum of process and device alternatives and provide improvement in parametric yield modeling in manufacturing. The integrated TCAD should address all phases of the product development life cycle: (a) research and development to explore the physical limits of processes and devices for technology development, (b) product integration to optimize process and device performance, and (c) manufacturing to control process variability, thereby increasing parametric yield. Integrated circuit (IC) design methodologies and the applied electronic design automation (EDA) tools are continuously changing. To keep pace with these changes, there is a critical need to acquire knowledge of available IC design tools, including EDA.

TCAD product suites have set a new standard for simulation. Tools are being designed to address the challenges by combining the best-in-class features together with a wide range of new features arising out of new technology generations. TCAD is predominantly physics based and has traditionally been the primary vehicle for predictive modeling of transistors and other active devices, considered to be part of front-end-of-line (FEOL) manufacturing. TCAD is also used to explore newer device designs and extrapolate to the next technology node, besides giving engineers a better understanding of the benefits and drawbacks of any modifications to existing manufacturing processes, as well as the development of compact/SPICE models.

Use of framework technology can encapsulate TCAD-specific knowledge from the user, transparently transfer data between simulation modules, and improve software modularity. TCAD environment (such as Synopsys and Silvaco) is developed by commercial vendors with highly user-friendly, interactive interfaces.

Typical requirements for the TCAD simulation graphical user interface (GUI) are:

- Easy to use. New users should be able to pick up the basics within minutes.
- Real-time help for the command syntax.
- Integration with other GUIs of the software suite, like the 3D setting-up tool and plotting GUI.

One highly demanding feature of the TCAD simulation GUI is the capability to perform batch simulation: this refers to a series of simulations associated with different process parameters, different device simulation parameters, etc. In the semiconductor industry this is useful because optimizing device performance often requires adjusting process parameters. There may be hundreds of possible process step combinations, so it is useful to automate the process as much as possible.

Process simulation creates a semiconductor device representation using modern semiconductor processing methods, such as diffusion, oxidation, etching, deposition, and implantation. Process and device simulation tools are linked to predict the influence of various IC processing steps on device and circuit performance. Most of the device simulators were designed to analyze device structures and solve the fundamental semiconductor equations, consisting of Poisson's equation and two carrier continuity equations, and solved numerically in 1D, 2D, and 3D space. The following is a brief introduction to the evolution of TCAD simulation tools where the role of CAD and the increasing dimension of simulation problems are highlighted. Only a brief history of the major development in the device and process simulation tools leading to commercial TCAD tools will be presented.

The seminal work of Gummel in 1964 led to the foundation of device simulation. Gummel's numerical approach was further developed and applied to simulate the p-n junction by De Mari in 1968. The discretization scheme reported for the transport equations by Scharfetter and Gummel is still almost universally used in device simulation. In the 1970s, 1D approaches were generally sufficient to deal with bipolar technology and early MOSFETs. Use of 1D charge transport phenomena was predominant in large-geometry devices. Extrapolation of quasi-2D doping distributions from sets of 1D profiles helped process and design optimization. During the 1970s, the finite element analysis of the semiconductor equations was reported, which helped in the development of the more general-purpose tools.

In the late 1970s, some of the first publicly available device simulation tools were released, viz., CADDETH from Hitachi, to simulate single-carrier field-effect transistor (FET) structures, SEDAN from Stanford University to simulate 1D bipolar devices, and the MOSFET simulation program MINIMOS from Vienna. During the 1970s and 1980s, several 1D and 2D programs were developed. Examples include SEDAN for 1D simulations and MINIMOS for 2D MOS transistor simulation, BAMBI for arbitrary semiconductor structures, and PISCES, a 2D finite element simulator which rapidly became an industry standard and formed the basis of future commercial products such as Silvaco-ATLAS. A nonplanar multidimensional device simulation tool for bipolar device analysis was FIELDAY from IBM. Other first-generation nonplanar device simulators include GEMINI and PISCES I from Stanford.

Since the 1980s, aggressive MOS scaling led to the very-large- and ultralarge-scale integration era based on CMOS technology. Full 2D simulators soon became indispensable to model, increasing process complexity and coupled physical effects, including local oxide isolation, dopant diffusion, subthreshold conduction, and parasitic phenomena such as latch-up and punch-through. Development of programs like PISCES II at Stanford University, DEVICE at AT&T Bell Laboratories, BAMBI at TU (Vienna), and HFIELDS at the University of Bologna came in the mid-1980s. PISCES II has been commercialized by Technology Modeling Associates (now Synopsys), and Silvaco and is the source of widely used device simulators MEDICI and ATLAS respectively. TMA developed its first 3D device simulator known as DAVINCI in 1991, which has been the basis of the Taurus 3D Device simulator from Synopsys.

With the increasing complexities of IC device structures due to continuous downscaling of feature size, 3D numerical analysis became critical. In 1987 TU, Vienna, announced the 3D device simulator MINIMOS Ver. 5 for MOSFET structures, silicon-on-insulator (SOI) transistors, and gallium arsenide metal–semiconductor field-effect transistors (MESFETs). MINIMOS supports transient analysis and MC modeling to replace the drift-diffusion approximation in critical device areas. Some of the other reported 3D device CAD tools include HFIELDS 3D, DESSIS, and FLOODS from the University of Florida. DESSIS is the basis of the commercial Sentaurus Device (SDevice) simulator from Synopsys.

In 1977, the first version of the 1D process simulator SUPREM developed in Stanford was released. Process simulation has continued to improve dramatically since the release of SUPREM. Since then, the process models in SUPREM have developed substantially and versions II and III of SUPREM were released in 1978 and 1983, respectively. In 1986, the most advanced 2D process simulation program SUPREM IV was released. The 2D SUPREM IV has been commercialized by Crosslight as CSUPREM, as ATHENA by Silvaco, and as TSUPREM4 by TMA. In 1992, ISE (now Synopsys) developed 1D and 2D process simulators TESIM and DIOS, respectively. DIOS is the basis of Sentaurus Process (SProcess) released by Synopsys in 2005. Other 3D process simulators include PROPHET from AT&T Bell Laboratories released in 1991.

2.2   Commercial TCAD Tools

The history of commercial TCAD began with the foundation of TMA in 1979. TCAD environments such as Synopsys [1–12] and Silvaco [13–23] have been developed by commercial vendors with highly user-friendly interactive interfaces. Currently, the main sources of commercial TCAD tools are Silvaco and Synopsys. Synopsys TCAD tools include Taurus-Process/Taurus-Device TSUPREM4/MEDICI for 2D TCAD and SProcess/SDevice for both 2D and 3D TCAD. Silvaco TCAD tools include ATHENA for 2D process simulation, ATLAS for 2D device simulation, and VictoryProcess/VictoryDevice for 3D simulation.

TCAD products should enable users to explore and optimize easily the broad spectrum of process and device alternatives and provide improvement in parametric yield modeling in manufacturing. Integrated TCAD should address all phases of the product development life cycle: (a) research and development to explore the physical limits of processes and devices for technology development, (b) product integration to optimize process and device performance, and (c) manufacturing to control process variability, thereby increasing parametric yield. IC design methodologies and the applied EDA tools are continuously changing. To keep pace with these changes, there is a critical need to acquire knowledge of available IC design tools, including EDA.

One of the most important GUIs for TCAD simulation is the plotting GUI. Without the ability to plot simulation results, there is not much point in doing the modeling in the first place. The plotting GUI allows the user to view both process and device simulation results: *IV* characteristics, bandgap diagrams, doping profiles, etc.

2.3 Silvaco Tool Overview

- ATLAS can be used in conjunction with the Virtual Wafer Framework (VWF) interactive tools. The VWF includes DeckBuild, Tonyplot, DevEdit, MaskViews, and Optimizer.
- ATHENA: 2D SSUPREM4-based process simulator.
- ATLAS: 2D (and basic 3D) device simulation.
- MaskViews: An IC Layout Editor.
- VictoryCell: GDS-based 3D process simulation.
- VictoryProcess: 3D process simulation.
- VictoryDevice: 3D device simulation.
- VictoryStress: 3D stress simulation.
- Virtual Wafer Fab: Wrapper of the different tool in a GUI.
- DeckBuild: Provides an interactive run time environment.
- TonyPlot: Visualization tool.
- TonyPlot 2D.
- TonyPlot 3D.
- Optimizer: Supports black-box optimization across multiple simulators.

Combining ATHENA, ATLAS, UTMOST, and SmartSpice makes it possible to predict the impact of process parameters on circuit characteristics. The VWF makes it convenient to perform highly automated simulation-based experimentation. It therefore links simulation very closely to technology development, resulting in significantly increased benefits from simulation use. The Silvaco simulation software suite may be used for modeling of III–V heterostructure devices in radio-frequency (RF)/microwave front-end electronics. The software can characterize all DC and RF data of the device. It is suitable for compound semiconductor–based solar/optical devices and may be used for process design, analysis, and optimization of various fabrication technologies. Silvaco TCAD offers complete and well-integrated simulation software for all aspects of solar cell technology. TCAD modules required for solar cell simulation include S-PISCES, Blaze, Luminous, TFT, Device3D, Luminous3D, and TFT3D. The TCAD-driven CAD approach provides the most accurate models to device engineers. Silvaco is a one-stop vendor for all companies interested in advanced solar cell technology simulation solutions.

2.3.1 MaskViews

IC Layout Editor is a versatile IC layout editor used to specify layout information to process simulators. MaskViews supports simulation-based experimentation with layout variations. Experimentation-based simulation was previously restricted to the varying of process flow parameters only. MaskViews supports experimentation dealing with phase shift masking technologies, critical dimensions, misalignment tolerances, and global shrinks. It is fully interfaced to GDS2 stream formats so that complete IC layouts may be imported and exported. Small subregions can be selected for detailed analysis. MaskViews can be used interactively, in which case it communicates via DeckBuild.

2.4 ATHENA

ATHENA is a 1D, 2D, and 3D process simulator used to model semiconductor devices. It is very effective tool, which replaces costly experiments on real-time fabrication of devices. Each fabrication step can be modeled and simulated in ATHENA, which includes oxidation, deposition, ion implantation, conventional furnace annealing, spike anneal, geometric etches, and lithography. ATHENA input and output profiles are shown in Fig. 2.3.

ATHENA simulation involves the following steps:

1. Generating an ATHENA input file
2. Running an ATHENA simulation
3. Analyzing an ATHENA output file

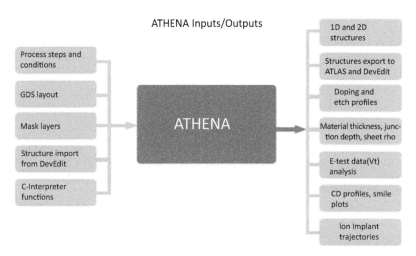

Figure 2.3 ATHENA input and output profiles.

Basic operations to create an ATHENA input file involves the following steps:

1. Developing a good simulation grid
2. Defining an initial substrate
3. Performing layer deposition
4. Performing geometrical etching
5. Performing ion implantation and diffusion
6. Specifying the electrodes
7. Saving the structure file

The simulation problem is defined using the following steps:

1. To initial geometry
2. To specify a simulation grid
3. To define an initial substrate
4. To specify process steps
5. To specify the electrodes
6. To save a structure file
7. To access the ATHENA coefficients file
8. To generate an EXTRACT statement

The sequence of process steps involves:

1. Epitaxial growth
2. Layers deposition
3. Geometrical etching
4. Ion implantation
5. Diffusion

Physical models needed in simulation:

- Implant models
- Diffusion models
- Oxidation models

ATHENA output file: Creating a device structure

The main ATHENA output is the standard structure file, a universal file format used by Silvaco simulation programs. The STRUCTURE statement of ATHENA creates a standard structure file (.str), which contains mesh and solution information, model information, and other related parameters. The details of export process which links the process to the device simulation depends on the software suite being used but the principles are usually similar. The saved structure file can be used for further simulation/visualization by:

- ATHENA to continue process simulation
- ATLAS or other device simulators to perform electrical analysis
- TonyPlot to graphically display the structure created by ATHENA
- DevEdit to modify the mesh and the structure before running a device simulation

The simulation grid represents the points (nodes) of the structure where the model equations are solved. Therefore the correct specification of a grid is critical in process simulation. The number of nodes in the grid has a direct influence on simulation accuracy and time. A finer grid should exist only in the critical areas of the simulation structure (where ion implantation will occur or where a p-n junction will be formed). The simulators are popular due to two main reasons. First, they help in predicting trends and understanding the device performance better. Second, the simulators are used to optimize a device structure for performance, which can then be verified by device fabrication. ATHENA features include:

- The ATHENA process simulation framework enables process and integration engineers to develop and optimize semiconductor manufacturing processes.
- ATHENA provides an easy-to-use, modular, and extensible platform for simulating ion implantation, diffusion, etch, deposition, lithography, oxidation, and silicidation of semiconductor materials.
- ATHENA replaces costly wafer experiments with simulations to deliver shorter development cycles and higher yields.
- SSuprem4 is the state-of-the-art 2D process simulator widely used in semiconductor industry for design, analysis, and optimization of Si, SiGe, and compound semiconductor technologies.
- SSuprem4 accurately simulates all major process steps using a wide range of advanced physical models for diffusion, implantation, oxidation, silicidation, and epitaxy.
- MC Implant is a physics-based 3D ion implantation simulator to model stopping and ranges in crystalline and amorphous materials.
- MC Implant accurately predicts implant profiles and damage for all major ion/target combinations.
- Elite is an advanced 2D moving boundary topography simulator for modeling physical etch, deposition, reflow, and chemical-mechanical polishing (CMP) planarization processes.
- MC Etch and Deposit is an advanced topology simulator.
- MC Etch and Deposit includes several Monte Carlo–based models for simulation of various etch and deposit processes which use a flux of atomic particles.

2.5 ATLAS

ATLAS is a physics-based device simulator. It provides general capabilities for 2D and 3D simulations of semiconductor devices. It specifies the device simulation problems by defining the physical structure to be simulated, the physical models to be used, and the bias conditions for which electrical characteristics are to be simulated (Fig. 2.4). The electrical characteristics predicted by ATLAS can be used as input by the UTMOST device characterization and SPICE modeling software. Compact models based on simulated device characteristics can then be supplied to circuit designers for groundwork circuit design. ATLAS features include:

- The ATLAS device simulation framework enables device technology engineers to simulate the electrical, optical, and thermal behavior of semiconductor devices.
- ATLAS provides a physics-based, easy-to-use, modular, and extensible platform to analyze DC, AC, and time domain responses for all semiconductor-based technologies in two and three dimensions.
- Device designs for simulation may be created directly within ATLAS, drawn in DevEdit, or imported from the ATHENA framework.

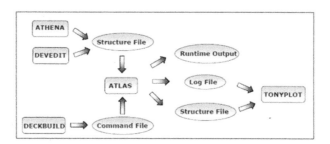

Figure 2.4   ATLAS inputs and outputs. Source: Maria Concetta Allia.

ATLAS is often used in conjunction with the ATHENA process simulator, which predicts the physical structures that result from processing steps. The resulting physical structures are used as input by ATLAS, which then predicts the electrical characteristics associated with specified bias conditions. The combination of ATHENA and ATLAS makes it possible to determine the impact of process parameters on device characteristics. ATLAS produces three types of output files. The first type of output file is the run-time output, which gives the progress and the error and warning messages as the simulation proceeds. The second type of output file is the log file, which stores all terminal voltages and currents from the device analysis. The third type of output file is the solution file, which stores

2D and 3D data relating to the values of solution variables within the device at a given bias point. Firstly, ATLAS helps in predicting trends and understanding the device performance better. Second, the simulators are used to optimize a device structure for performance, which can then be verified by device fabrication.

2.5.1 Physical Structure

A device structure can be defined in several different ways in ATLAS. A fabrication simulator ATHENA can be used to "build" a device by simulating the device's fabrication or a CAD tool like DEVEDIT can be used to define the structure. This device structure can then be read into DeckBuild, which is the run-time environment of ATLAS. Another way of defining the structure is to directly construct the device using the ATLAS command language. This method was used in defining the SiGe heterojunction bipolar transistor (HBT) structure for this power amplification investigation. To define a device through the ATLAS command language, a simulation domain is first defined by defining a mesh and setting air as the material for this domain. The regions within the domain are then defined by defining the size of that region and the type of material. Defining a material automatically overrides the default material (air) set in the first step. After the regions are defined, the locations of the electrodes are specified. The final step is to specify the doping in each region.

Defining the mesh is one of the most important steps of the structure definition. By increasing the number of mesh points, the accuracy of the result of the simulation can be increased. This is because each mesh point represents a point in the structure where the physical models are solved. But at the same time, increasing the mesh points increases the simulation time. Moreover, the number of mesh points cannot exceed a particular limit as dictated by the simulator. Therefore, there is always a trade-off between the accuracy and the simulation time. For getting the best results, the mesh density is large at critical regions of the structure, like the junction of the emitter and the base, while keeping the mesh density to a minimum in regions that are less critical. It is also important, especially in the case of comparative study, to avoid changing mesh density. When using this tool to define a structure, the information described in the following four subsections must be specified in the order listed. In the following, we describe each of these in more detail. Advantages of ATHENA/ATLAS are:

- 3D simulation built in
- Seamless transition from 2D to 3D
- Excellent user interface
- Parallel 3D solver (takes advantage of a modern multicore CPU)
- Adaptive meshing and clever 3D meshing algorithm

2.5.2 Structure Editing

To prepare a structure for device simulation, structure editing is performed using Sentaurus Structure Editor. The n-type metal-oxide-semiconductor (NMOS) structure from process simulation is reduced in size by removing the lower portion of the substrate. Next, contacts are placed for the substrate, gate, source, and drain. After contact placement, the structure is ready to be remeshed. Since mesh requirements are different for process and device simulations, the structure editor is also utilized to remesh the structure prior to device simulation.

An optimum mesh is important to resolve key regions of the transistor pertaining to charge injection and clock feed-through; these areas include the inversion layer and overlap regions. To properly resolve the inversion layer, the mesh in the gate oxide is 10 Å, followed by a fine mesh in the first 300 Å of the channel. In the channel's first 10 Å the mesh spacing is 2 Å, and as the channel depth increases to 300 Å the mesh spacing increases to 20 Å. The source and drain are remeshed to complete the NMOS device.

2.5.3 Meshing

TCAD simulators are based on finite element solutions to solve the nonlinear partial differential equations (PDEs). Therefore, good meshing techniques are required to perform numerically stable and accurate simulations. Automated mesh generation is a major challenge for semiconductor device and process simulation. Various mesh generation schemes have been proposed and implemented to solve systems of linear and nonlinear PDEs. Optimum meshing is important to maintain accuracy without dramatically increasing process simulation time. In a transistor, an initial mesh is defined by specifying the spacing in the vertical direction. Additional mesh refinements are necessary in critical areas of the device, such as the channel, source, drain, and lightly doped drain (LDD) regions.

2.5.4 Mesh Definition

The difficulties in mesh generation are a challenging task in TCAD and call for automatic and adaptive grid generation techniques. Mesh definition is important for simulation accuracy and time. A fine mesh gives an accurate solution but increases simulation time. Device simulation involves the solution of coupled PDEs which describe the evolution of either geometry or impurity distribution as a result of manufacturing process steps, or internal physical quantities in response to electrical boundary conditions. Solutions of coupled PDEs can only obtained numerically; thus, a proper discretization (mesh generation) procedure is required. Mesh generation has thus a crucial impact on the convergence, accuracy, and efficiency of the simulation. Also, meshing has become a major issue because device architectures are now essentially 3D. Therefore, automatic grid generation and adaptation are highly desirable, for improving both the trade-off between computational complexity and solution accuracy. To optimizing the mesh grid is the important goal for MOSFET structures. Automated gridding procedures (adaptive meshing) are desirable in process and device simulations. A mesh should be refined in the key areas. These areas are:

- Around junctions and depletion regions
- Inversion regions
- Areas of high electric field
- Areas of current flow

2.5.5 Regions and Materials

MOS devices have oxide, poly, gate contact, source contact, drain contact, and substrate regions. Region command specifies the location of materials in a previously defined mesh. Every triangle must be defined as a material. Material parameters required for device simulations are:

- Lattice constant
- Bandgap
- Density of states
- Bandgap narrowing
- Intrinsic carrier concentration
- Dielectric constants
- Electron affinity
- Mobility

2.5.6 Physical Models

Physical models are a set of mathematical equations used to define the physical behavior of a system. The basic physical models used to define the device physics simulated here are derived from Maxwell's equations. These include Poisson's equation, current continuity equations, and carrier transport equations. An electric field accelerates electrons and holes. But eventually they lose momentum as a result of various scattering processes. These scattering mechanisms are lattice vibrations (phonons), impurity ions, other carriers, surfaces, and other material imperfections. Therefore mobility is the function of the local electric field, lattice temperature, doping concentration.

Mobility modeling is divided into:

- Low-field mobility
  - o Constant low-field mobility
  - o Concentration-dependent low-field mobility
  - o Analytic low-field mobility
  - o Arora model for low-field mobility
  - o Carrier–carrier scattering model for low-field mobility
  - o Klaassen's unified low-field mobility
- Inversion layer mobility
  - o Lombardi CVT model
  - o Extended CVT model
  - o Yamaguchi model

    o Tasch model
- Perpendicular electric field–dependent mobility
  - o Watt model
  - o Shirahata mobility model
- Parallel electric field–dependent mobility
- Carrier temperature–dependent mobility
- Carrier generation-recombination models

Carrier generation-recombination is the process through which the semiconductor material attempts to return to equilibrium after being disturbed from it. The reasons that cause generation-recombination are:

- Phonon transitions
- Photon transitions
- Auger transitions
- Surface recombination
- Impact ionization
- Tunneling

2.5.6.1 Models

- Shockley–Read–Hall (SRH) recombination model
- SRH concentration-dependent lifetime model
- Klaassen's concentration-dependent lifetime model
- Trap-assisted tunneling
- Radiative recombination (photon generation)
- Auger recombination
- Standard Auger model
- Klaassen's temperature-dependent Auger model
- Narrow-bandgap Auger model
- Surface recombination

2.5.7 Impact Ionization Models

- Local electric field models for impact ionization
  - o Selberherr impact ionization model
  - o Valdinoci impact ionization model

o Grant's impact ionization model
- Nonlocal carrier energy models for impact ionization
    o Concannon impact ionization
    o Band-to-band tunneling

2.5.7.1 C-Interpreter functions

The list of models available via the C-Interpreter includes all basic physical models: carrier velocities and mobilities, temperature- and composition-dependent band parameters, position-dependent composition, recombination models and their parameters, materials parameters as a function of composition, interface parameters, etc.

2.5.8 Gate Current Models

The conductance of the insulating film would ideally be considered as zero. But, for the sub-0.5 pm generation of MOS devices there is now considerable conductance being measured on the gate contacts. The negative side of this gate current is responsible for the degradation in device operating characteristics with time. The models are:

- Fowler–Nordheim tunneling
- Lucky electron hot-carrier injection model
- Concannon's injection model

2.5.9 Bandgap Narrowing

As the doping level increases, a decrease in the bandgap separation occurs, where the conduction band is lowered by approximately the same amount as the valence band is raised.

2.5.10 Solution Methods

- Numerical Methods
    o Newton's method
    o Gummel's method
    o Discretization
    o Continuation method
- Electrical analyses
    o DC analysis

o Transient analysis
o Small-signal analysis
o Impact ionization analysis
o Gate current analysis

ATLAS and ATHENA solve differential equations describing device fabrication and electrical properties using numerical techniques. These equations are solved in two dimensions, which is often difficult to do analytically. Therefore having a computer program to solve and display the results of 2D numerical simulation is an invaluable part of modern day IC manufacturing. We shall not discuss how ATLAS and ATHENA solve differential equations, other than to say that a finite element method is used to solve difference equations on a grid of point, known as a mesh. The mesh has a triangular geometry, and ATLAS and ATHENA calculate average quantities within the triangles. A complete description with respective syntaxes is given in the ATHENA and ATLAS manuals from Silvaco International.

The trend toward using laptops for mobile computing shows a steady increase in the demand for Windows versions of TCAD tools. In response, Silvaco supports the following ATHENA and ATLAS tools for TCAD for Windows. Products available in PC-TCAD from Silvaco are shown in Figs. 2.5 and 2.6.

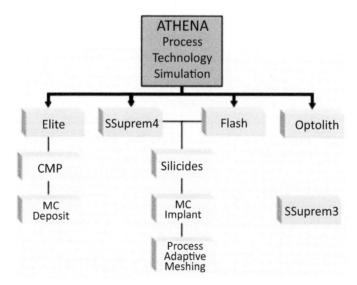

Figure 2.5 PC version of the ATHENA module.

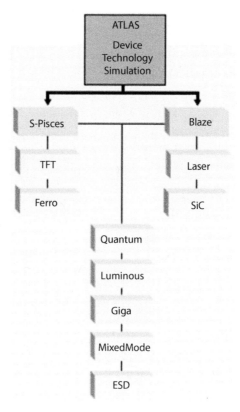

Figure 2.6 PC version of the ATLAS module.

2.5.11 VictoryCell

VictoryCell is a layout-driven 3D process simulator specifically designed for large-scale structures for creating 3D devices. VictoryCell is capable of modeling etch, deposit, implant, diffusion and photolithography. It is user-friendly and uses SUPREM-like syntax. It can be interfaced to both Silvaco's 3D device simulators, such as Device-3D and VictoryDevice. Using VictoryCell one can create a layout-driven device from either GDSII or Silvaco layout formats. Layout-driven simulation allows creation of high-aspect-ratio 3D structures. VictoryCell is equivalent to Synopsys SProcess and T-SUPREM4 (SUPREM IV). VictoryCell is a 3D process simulator designed to be a very fast way of creating devices. The choice of mesh formats that can be output is optimized to be as device simulator friendly as possible, minimizing device simulation times. VictoryCell may be used for simulation of numerous technologies, such as SiC, IGBT, and MOSFET. The VictoryCell information flow is shown in Fig. 2.7.

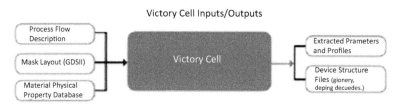

Figure 2.7 VictoryCell information flow.

To perform a 3D simulation, one needs the mask-layout for all the process steps. The mask layout can be designed in a variety of drafting tools and specialized EDA tools like Cadence Virtuoso and Tanner L-Edit. These tools are used to create a GDSII file which is the current industry standard for IC layout work. Silvaco MaskEditor may be used to import GDSII files. If GDSII files are not available, MaskEditor may be used to create the geometric shapes and layers used in the process simulation. The 3D process simulator takes the input files created by MaskEditor and performs 3D process simulation and export the device structure to the device simulator. The device simulator uses the exported mesh and material information to perform electrical, thermal and, optical simulations.

2.5.12   VictoryProcess

The 3D process simulator VictoryProcess allows the user to generate a wide range structures in order to obtain the device of the desired shape. Most of the supported operations correspond to real technological processes (etching or deposition, CMP, epitaxy, and others) and one can establish a direct link between the technological processes step and an input deck statement of VictoryProcess. The VictoryProcess information flow is shown in Fig. 2.8. It can simulate complicated full physics-based etching and deposition processes which take into account

- Reactor characteristics (particle flux)
- Shading effects and/or secondary effects like redeposition of the etched material

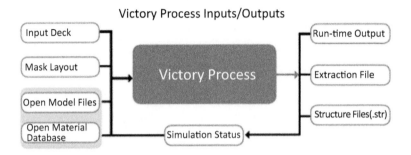

Figure 2.8 VictoryProcess information flow.

As full physics-based simulations can take a long time and may be influenced by numerical error. Using either the geometrical mode of VictoryProcess or very simple physical models, many advanced operations may be performed very quickly and accurately.

VictoryProcess features include:

- Fast 3D structure prototyping capability
- Comprehensive set of diffusion models: Fermi, fullcpl, single-pair, and five-stream
- Physical oxidation simulation with stress analysis
- Accurate and fast Monte Carlo implant simulation
- Efficient multithreading of time critical operations of Monte Carlo implantation, diffusion, oxidation, and physical etching and deposition
- Sophisticated multiparticle flux models for physical deposition and etching with substrate material redeposition
- Open architecture allowing easy introduction and modification of customer-specific physical models
- Seamless link to 3D device simulators, including structure mirroring, adaptive doping refinement, and electrode specification
- Easy-to-learn and user-friendly SUPREM-like syntax
- ATHENA compatibility
- Convenient calibration platform
- Automatic switching from 1D, 2D, and 3D modes

2.5.13   VictoryStress

The effect of mechanical stress in semiconductors was reported by Smith [24] who studied the relationship between resistivity change and tension in silicon and germanium in 1954, establishing a linear dependence. Accurate simulation of mechanical stress/strain distributions generated during device fabrication is an important part of technology and device design. In many cases, stress effects should be taken into account to predict better manufacturability and to increase reliability of semiconductor devices.

In recent years, the stress simulation has become a critical issue in TCAD due to advances in stress engineering. Stress engineering could be described as a collection of device optimization methods based on deliberate introduction of stresses into the device cell structure. These methods include dual stress liners, strain SiGe channels, SiGe (SiC) S/D pockets, hybrid orientation of NMOS and p-type metal-oxide-semiconductor (PMOS) devices, and etch stop layers. All these methods are used to improve device performance by altering carrier mobility in specific areas of the device structure. VictoryStress is a generic 3D stress simulator that allows accurate prediction of stresses generated during semiconductor fabrication, as well as assisting users in all aspects of stress engineering. The three main modules in VictoryStress are input file generation, 3D VictoryStress simulation, and output generation. The VictoryStress information flow is shown in Fig. 2.9.

Figure 2.9   VictoryStress information flow.

2.5.13.1   VictoryStress features and capabilities

VictoryStress provides a comprehensive set of models and capabilities covering various aspects of stress simulation and stress engineering:

- Layout-driven stress analysis
- Comprehensive material stress models, including the dependence of elasticity coefficients on crystal orientation
- Generic 3D anisotropic stress simulation for crystalline materials suitable for a variety of stress analysis such as:
  - Strained PMOS, compressive capping layers
  - Mobility enhancements in fin field-effect transistor (finFET) channels
- Stress analysis of arbitrary 3D device structure
- Import of device structures from VictoryProcess, VictoryCell, and ATLAS3D
- Export of 3D stress distributions to VictoryDevice and ATLAS3D
- Models for various sources of strain and stress
- Thermal mismatch between material layers
- Local lattice mismatch due to doping
- Initial intrinsic stress in specified regions
- Hydrostatic stress from capping layers
- Stress/strain generated in previous processing step (e.g., oxidation)
- Stress simulation for various crystalline (e.g., Si, SiGe, GaAs) and isotropic (e.g., silicon nitride and oxide) materials
- Generic 3D anisotropic stress simulation accounts for wafer orientation and arbitrary wafer flat rotation
- Estimation of mobility enhancement factors (p- and n-type) by use of piezoresistivity model devices
- Ability to be used with Virtual Wafer Fab for design of experiments to analyze stress dependence on process and geometrical parameters of semiconductor

2.5.14 VictoryDevice

VictoryDevice provides general capabilities for physics-based 2D and 3D simulation of semiconductor devices. VictoryDevice is designed to be used with the VWF interactive tools. The VWF interactive tools are DeckBuild, TonyPlot, TonyPlot3D, DevEdit, MaskViews, and Optimizer. The VictoryDevice information flow is shown in Fig. 2.10. VictoryDevice provides a wide set of physical models, including:

- DC, small-signal AC, and fully time-dependent simulations
- Drift-diffusion and energy balance transport
- Lattice heating and heat sinks
- Fermi–Dirac and Maxwell–Boltzmann statistics
- Bulk and field-dependent carrier mobilities
- Heavy doping effects
- Ohmic and Schottky contacts
- SRH, radiative, Auger, and surface recombination
- Impact ionization
- Band-to-band tunneling
- Thermionic emission currents

VictoryDevice uses powerful numerical techniques, including:

- Accurate and robust discretization techniques
- Gummel, Newton, and block-Newton nonlinear iteration strategies
- Efficient direct and iterative solvers
- Powerful initial guess strategies
- Small-signal calculation techniques that converge at all frequencies
- Stable and accurate time integration

This section provides the basic equations of elasticity and the solution approaches to determine the stress state. Apart from process-induced stress, layout-dependent stress effects contribute to performance variations in ICs. In earlier technologies, transistor sizes were large enough such that their electrical behavior was independent of the final layout. However, in highly scaled technologies with smaller geometries, the electrical performance of a transistor has become increasingly dependent on its location in the layout. Unwanted mechanical stresses in the layout, as well as unwanted variations on-chip stresses, affect transistor electrical properties due to piezoresistivity and electronic band deformation. Thus stress effects significantly affect design methodologies in modern ICs.

In addition, advanced packaging techniques have further resulted in proliferation of unwanted sources of mechanical stress, and the variations caused by mechanical stress effects have become comparable to those from lithography variations. Thus, it is important to consider the mechanical stress effects early in the design. Since we are mainly concerned with stress and strain, it is essential to understand the basics of engineering mechanics like stress, strain, and mechanical properties. Within the elastic limit, the property of a solid material to deform under the application of an external force and to regain its original shape after the force is removed is referred to as its elasticity. It is the law of Hooke, who described the elastic relation between the mechanical constraint and deformation which a material will undergo. The external force applied on a specified area is known as stress, while the amount of deformation is called the strain. In this section, the theory of stress, strain, and their interdependence is briefly discussed. Following the ATLAS manual, we describe sources of strain/stress and material parameters used for stress calculations in isotropic and anisotropic (crystalline) materials. Finally, we will discuss the simulation procedure and numerical methods used in VictoryStress.

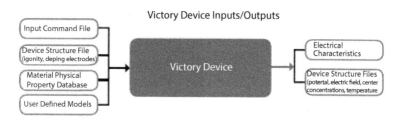

Figure 2.10 VictoryDevice information flow.

2.6   Stress Modeling

The stress (σ) at a point may be determined by considering a small element of the body enclosed by area (ΔA), on which forces act (ΔP). Its unit is Pascal (Pa). By making the element infinitesimally small, the stress (σ) vector is defined as the limit

$$\sigma = \lim_{\Delta A \to 0} \frac{\Delta F}{\Delta A} = \frac{dF}{dA}$$

From Fig. 2.11 one can observe that the force acting on a plane can be decomposed into a force within the plane, the shear components, and one force perpendicular to the plane, the normal component.

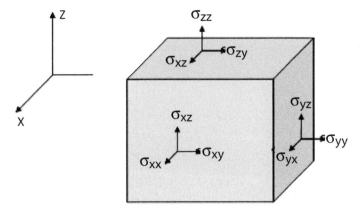

Figure 2.11 Stress components acting on infinitesimal cube.

The shear stress can be further decomposed into two orthogonal force components, giving rise to three total stress components acting on each plane. Figure 2.11 shows normal and shears stresses in the X, Y, and Z directions, acting on different planes of the cube. The first subscript identifies the face on which the stress is acting, and the second subscript identifies the direction. The σ_{ii} components are the normal stresses, while the σ_{ij} components are the shear stresses.

2.6.1 Stress–Strain Relationship

Generally, the stress tensor σ is symmetric 3×3 matrices. Therefore, it only has six independent components. With the index transformation rule, it can be written in a six-component vector notation $\sigma_{11} \rightarrow \sigma_1$, $\sigma_{22} \rightarrow \sigma_2$, $\sigma_{33} \rightarrow \sigma_3$, $\sigma_{23} \rightarrow (\sigma_4)/2$, $\sigma_{13} \rightarrow (\sigma_5)/2$, $\sigma_{12} \rightarrow (\sigma_6)/2$ that simplifies tensor expressions. For example, to compute the strain tensor (which is needed for the deformation potential model), the generalized Hooke's law for anisotropic materials is applied:

$$\varepsilon_{ij} = \sum_{j=1}^{6} S_{ij}\sigma_j$$

where S_{ij} is the elasticity modulus. In crystals with cubic symmetry, such as silicon, the number of independent coefficients of the elasticity tensor (as other material property tensors) reduces to three by rotating the coordinate system parallel to the high-symmetric axes of the crystal. This gives the following elasticity tensor $\bar{S}$:

$$\bar{S} = \begin{bmatrix} S_{11} & S_{12} & S_{12} & 0 & 0 & 0 \\ S_{12} & S_{11} & S_{12} & 0 & 0 & 0 \\ S_{12} & S_{12} & S_{11} & 0 & 0 & 0 \\ 0 & 0 & 0 & S_{44} & 0 & 0 \\ 0 & 0 & 0 & 0 & S_{44} & 0 \\ 0 & 0 & 0 & 0 & 0 & S_{44} \end{bmatrix}$$

where the coefficients S_{11}, S_{12}, and S_{44} correspond to the parallel, perpendicular, and shear component, respectively.

The stiffness tensor a function of Young's modulus E and Poisson's ratio v of the material. The nonzero components are given as

C1111 = C2222 = C3333 = $E(1 - v)/(1 + v)(1 - 2v)$

C1122 = C2233 = C1133 = $Ev/(1 + v)(1 - 2v)$

C2211 = C3322 = C3311 = $Ev/(1 + v)(1 - 2v)$

C1212 = C3131 = C2323 = $E/2(1 + v)$

Mechanical stress causes change in the band edges in silicon. These band edge shifts are given by deformation potential theory. The shifts for the conduction band edges are given by

$$\Delta E_c^{(i)} = D.\text{DEFPOT}(\varepsilon_{xx} + \varepsilon_{yy} + \varepsilon_{zz}) + U.\text{DEFPOT} * \varepsilon_{ii}$$

where $\Delta E_c^{(i)}$ is the shift in the band edge of the i-th ellipsoidal conduction band minima. The parameters $U.\text{DEFPOT}$ and $D.\text{DEFPOT}$ are the user-definable dialation and shear deformation potentials, respectively, for the conduction band.

The ε_{xx}, ε_{yy}, and ε_{zz} parameters are the diagonal components of the strain tensor. The shifts in the valence band edges are calculated by

$$\Delta E_v^{(hl)} = A.\text{DEFPOT}(\varepsilon_{xx} + \varepsilon_{yy} + \varepsilon_{zz}) \pm \sqrt{\xi}$$

where $\Delta E_v^{(hl)}$ are the band edge shifts in the light and heavy hole valence band maxima. The ξ parameter is given by

$$\xi = \frac{1}{2} B.\text{DEFPOT}^2 \left[(\varepsilon_{xx} - \varepsilon_{yy})^2 + (\varepsilon_{yy} - \varepsilon_{zz})^2 + (\varepsilon_{zz} - \varepsilon_{xx})^2 \right]$$
$$+ C.\text{DEFPOT}^2 \left(\varepsilon_{xy}^2 + \varepsilon_{yz}^2 + \varepsilon_{zx}^2 \right)$$

where ε_{xy}, ε_{yz}, and ε_{zz} are the off-diagonal components of the strain tensor.

The strain components are calculated from the stress components stored in the input structure file if included (typically imported from ATHENA). The conversion between stress and strain is given by

$$\varepsilon_{xx} = \sigma_{xx} s_{11} + \sigma_{yy} s_{12}$$

$$\varepsilon_{yy} = \sigma_{xx} s_{12} + \sigma_{yy} s_{11}$$

$$\varepsilon_{zz} = 2\sigma_{yy} s_{44}$$

Here, σ_{xx}, σ_{yy}, and σ_{zz} are the diagonal components of the stress tensor and $s11$, $s12$, and $s44$ are the material compliance coefficients.

The compliance coefficients are given by

$$s_{11} = \frac{c_{11} + c_{12}}{c_{11}^2 + c_{11}c_{12} - 2c_{12}^2}$$

$$s_{12} = \frac{-c_{12}}{c_{11}^2 + c_{11}c_{12} - 2c_{12}^2}$$

$$s_{44} = \frac{1}{c_{44}}$$

where c_{11}, c_{12}, and c_{44} are the elastic stiffness coefficients. The stiffness coefficients for silicon and germanium are given by

$$\left.\begin{array}{l} c_{11} = 163.8 - T \times 0.0128 \\ c_{12} = 59.2 - T \times 0.0048 \\ c_{44} = 81.7 - T \times 0.0059 \end{array}\right\} \text{Si}$$

$$\left.\begin{array}{l} c_{11} = 126.0 \\ c_{12} = 44.0 \\ c_{44} = 67.7 \end{array}\right\} \text{Ge}$$

where T is the temperature. The stiffness coefficients for SiGe at composition are given by linear interpolation.

If the stress tensor is not loaded from a structure file, one can specify the values of the strain tensor. If the stress tensor is not loaded from a structure file and the strain tensor is not specified, then the strain tensor is calculated as

$$\varepsilon_{xx} = \varepsilon_{yy} = \frac{(1+v)}{(1-v)} \frac{a_{\text{SiGe}} - a_{\text{Si}}}{a_{\text{SiGe}}}$$

where v is Poisson's ratio, a_{SiGe} is the lattice constant of SiGe, and a_{Si} is the lattice constant of Ge. The lattice constants and Poisson's ratio for Si and Ge are given by:

$$a_{Si} = 5.43102 + 1.41 \times 10^{-5}(T - 3000)$$

$$a_{Ge} = 5.6579 + 3.34 \times 10^{-5}(T - 3000)$$

$$v_{Si} = 0.28$$

$$v_{Ge} = 0.273$$

The lattice constant and Poisson's ratio for $Si_{1-x}Ge_x$ is calculated by linear interpolation. The net changes in the band edges, under Boltzmann's statistics, are given by

$$\Delta E_c = kT \ln \left[\sum_{=1}^{3} \frac{\exp\left(-\dfrac{\Delta E_c^{(i)}}{kT}\right)}{3} \right]$$

$$\Delta E_v = k_T \ln \left[\frac{r}{1+r} \exp\left[\frac{\Delta E_v^{(1)}}{kT}\right] + \frac{1}{1+r} \exp\left[-\frac{\Delta E_v^{(h)}}{kT}\right] \right]$$

Here, the parameter r is given by

$$r = (m_1/m_h)^{3/2}$$

where m_l and m_h are the effective masses of light and heavy holes, respectively. To enable the model for the stress-dependent bandgap in silicon, one needs to specify the STRESS parameter in the MODELS statement.

2.6.2 Mobility

Although the mechanisms behind the enhancement in stress-/strain-induced mobility are fairly well understood qualitatively, quantitative evaluation is much more difficult as the type of mechanical stress induced is indeed very varied [25]. It may be: uniaxial stress or biaxial, with local inhomogeneities, with effects which may be very different depending on crystal orientations and also the type of carriers (electrons or holes). The switching speed of an ideal transistor can be increased primarily by two ways, physical gate length scaling and carrier mobility enhancement. In strained-silicon technology, the switching speed is enhanced solely by enhancing the carrier mobility. The carrier mobility is given by

$$\mu = q\frac{\tau}{m^*}$$

where $1/\tau$ is the scattering rate and m^* is theconductivity effective mass. The carrier mobility is enhanced by strain by reducing the effective mass and/or the scattering rate. Electron mobility is enhanced by both the phenomena, while for holes only mass change due to band warping is known to play a significant role at the current stress levels in production. The mobility is directly related to the carrier velocity υ and applied external electric field $\mathbf{E}$ by

$$\upsilon = \mu.E$$

The impact of, for example, acoustic and optical phonons, impurities, and surface roughness on the carrier mobility at a low electric field is generally simulated using several models. Each model calculates a proper mobility, depending on the effect. The different mobility factors are finally combined to one low-field mobility μ_{low} by Mathiessen's rule:

$$\frac{1}{\mu_{low}} = \frac{1}{\mu_{b1}} + \frac{1}{\mu_{b2}} + \cdots + \frac{1}{\mu_{s1}} + \frac{1}{\mu_{s2}} + \cdots$$

where μ_{bi} are the bulk mobility contributions and μ_{si} are the surface mobility contributions.

$$\mu_{const} = \mu_L \left(\frac{T}{300 \text{ K}}\right)^{-\varsigma}$$

where ς is a fitting exponent which is typically 2.5 and 2.2 for electrons and holes, respectively. One contribution to the total carrier mobility, calculated by the simulation software, is the so-called constant mobility μ_{const}, which already includes the mobility reduction due to optical phonons and a temperature dependence.

However, at high electric fields ($>10^6$ V/cm) a saturation of the carrier velocity is observed in silicon. The saturation of the carrier drift velocity can be explained by the increasing probability of scattering events between carriers and phonons or impurities. To model this effect, the Caughey–Thomas model is used. This model assumes a velocity convergence to a fixed saturation velocity $\mathbf{V}_{sat}$ at infinitely high electric fields, which cannot be exceeded:

$$\mu_{high} = \frac{\mu_{low}}{\left[1 + \left(\dfrac{\mu_{low} F}{v_{sat}}\right)^{\beta}\right]^{1/\beta}}$$

The parameter β in the Caughey–Thomas model is a fit parameter, which is 1.1 for electrons and 1.2 for holes. For electrons in silicon, V_{sat} is 1.0×10^7 cm/s and for holes 0.85×10^7 cm/s.

2.6.3 SmartSpice

SmartSpice is a circuit design tool which is used for the general design purposes for circuit simulation of nonlinear DC, nonlinear transient, and linear AC analyses. SmartSpice simulates a circuit by calculating the behavior of all circuit components simultaneously. Through its many reliable models, SmartSpice bases its simulation on physical properties, as well as electrical parameters, to simulate the behavior of complex circuits. It simulates circuits and subcircuits consisting of the five most common semiconductor devices—diodes, junction gate field-effect transistors (JFETs), bipolar junction transistors (BJTs), MOSFETs, and MESFETs—and in addition it simulates resistors, capacitors, inductors, dependent and independent voltage and current sources, switches, and transition lines.

2.7 Synopsys TCAD Platforms

Synopsys is the worldwide leading provider of physics-based TCAD simulation software and services that enable semiconductor companies to save time and resources during IC product development. Sentaurus is the new-generation TCAD simulation platform from Synopsys. Sentaurus includes a comprehensive suite of core TCAD products for multidimensional process, device, and system simulations, embedded into a powerful user interface. The application space of Sentaurus spans the complete range of semiconductor technologies, from deep-submicron logic, memory, and mixed signals to smart power, sensors, compound semiconductors, optoelectronics, and RF. SProcess is a process simulator capable of handling 3D structures. SProcess is based on the FLOOPS program package, which was developed at the University of

Florida. The methodology for the simulation of process technologies and the command commands are very similar to those of Tsuprem4. The program can be controlled manually or through a command file. The command file is an ASCII text file, in which all necessary for the simulation commands are included.

Tsuprem4 is an easy-to-handle process simulator, with the 1D and 2D structure-handling capability. The grid is self-adapting, which means that it can automatically adapt changed conditions. For example, when the dopant distribution changes rapidly, the grid becomes finer in order to better resolve the dopant gradient. The Sentaurus TCAD suite bridges the needs of development and manufacturing engineers by improving semiconductor process control in manufacturing. It includes 2D and 3D simulation products that can be used to optimize new technologies and explore a broad range of process and device alternatives, and provides a mechanism to improve parametric yield in manufacturing. TCAD simulation results provide early feedback describing the impact of process and device design changes at any stage of the product development cycle. Sentaurus combines advanced, calibrated physical models, robust algorithms, and numeric and efficient meshing and structure-editing capabilities to generate accurate and predictive simulation results for a broad range of applications, including CMOS, bipolar, heterojunction, compound, power, memory, optoelectronics, analog/RF, and laser. Figure 2.12 shows a sample state-of-the-art tool chain from Synopsys.

Sentaurus Structure Editor is a CAD program for drawing simulation structures. It offers both graphical and command line interfaces to define geometries and analytical doping profiles. The drawing tools are fully featured, including primitive geometries, Boolean operations, and 3D extrusion and sweeping options. Sentaurus Structure Editor is also capable of defining meshing strategies based on interface, doping, and analytical schemes. Since the simulated FETs were generated from process simulation in SProcess, this tool is used to redefine the meshing file for more efficient device simulation.

SProcess of Synopsys is used for the process simulation. The software is an advanced 1D, 2D, and 3D process simulator which is suitable for silicon and other semiconductor materials. It includes several models for standard silicon processing steps, like oxidation, deposition, etching, implantation, diffusion, and silicidation processes. It also allows for designation of masks for processing steps to create patterned structures. Material parameters are taken from a parameter database developed by Synopsys. Furthermore, it accounts for mechanical stress, which has become a big topic in the CMOS industry in recent years.

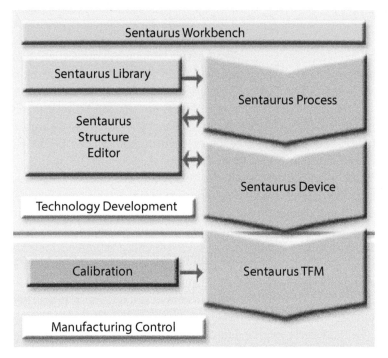

Figure 2.12 Sentaurus Workbench. Source: Sentaurus manual.

SDevice is a physics-based device simulation tool. By combining advanced physical models with robust numerical methods, it is capable of simulating the electrical characteristics of a wide range of semiconductor devices. The physical models implemented by SDevice must be chosen with care and an understanding of the models and the device being simulated. The semiconductor device is discretized by the points defined in the simulation mesh, and physical device equations are solved at each mesh node. It is capable of simulating isolated devices and devices connected in a circuit. Devices can be simulated under DC conditions for the *I–V* characteristics or in a mixed mode, allowing for an AC signal to be coupled to the system to measure the *C–V* characteristics of the simulated device. Sentaurus features include:

- Sentaurus Workbench, a flexible framework environment with advanced visualization and programmability
- SProcess, a 2D/3D process simulator
- SDevice, a 2D/3D device simulator
- Sentaurus Library, SIMS profiles, and calibrated model parameters for the latest technologies
- Sentaurus Structure Editor, a 2D/3D device editor with process emulation mode
- Sentaurus TFM, a process compact model extractor and optimizer
- Ligament
- Sentaurus Structure Editor
- Mesh and Noffset3D
- Tecplot SV
- Inspect
- Calibration Kit

Process-aware design for manufacturing tools from Synopsys includes:

- SProcess
- SDevice
- PCM Studio
- Raphael
- Seismos
- Fammos
- Paramos

2.7.1 Taurus-Device

Taurus-Device is a multidimensional device simulation tool. Taurus-Device simulates electrical and thermal characteristics of any semiconductor device in one, two, or three dimensions. A wide variety of devices, ranging from deep-submicron MOSFETs or bipolar devices to large-power device structures, can be modeled.

2.7.2 Taurus-Process

Taurus-Process is a multidimensional process simulation tool. Taurus-Process simulates all important fabrication steps used to manufacture semiconductor devices. It can simulate for a full 3D structure or in one or two dimensions. Typical processes simulated are deposition, etching, ion implantation, diffusion, and oxidation. Taurus-Process has a direct interface to Taurus-Device for electrical and thermal device characterization.

2.7.3 Device Simulation

SDevice provides 1D, 2D, and 3D simulations for a wide range of semiconductor devices. It features advanced simulation models for charge carrier transport in semiconductors, especially those needed for deeply nanoscaled CMOS devices with high numerical robustness. SDevice offers three different models of charge carrier transport in silicon: drift-diffusion, hydrodynamic transport, and Monte Carlo transport. Mixed-mode simulations, which can be understood as SPICE simulations with numerical device models, offer the ability to investigate the electrical behavior of novel CMOS devices under circuit conditions. The major device simulations are done by conventional drift-diffusion simulations, since they are fast and ensure a reliable convergence behavior and many models for different physical effects are already available for this kind of simulation method.

The calculation of the current densities of electrons J_n and holes J_p is dependent on the respective simulation method that is used. In the drift-diffusion approach, the electron current density J_n is computed by solving the relations

$$J_n = qn\mu_n F + qD_n \frac{dn}{dx}$$

and the hole current density is calculated by

$$J_p = qn\mu_p F - qD_p \frac{dp}{dx}$$

where μ_n and μ_p are the electron and hole mobilities, F is the electric field, which is calculated from the electrostatic potential, and D is the diffusion coefficient of either electrons (D_n) or holes (D_p).

2.7.4   Carrier Recombination-Generation

Generation of a charge carrier occurs if energy is applied to a valence electron, higher than the bandgap of the semiconductor material ($E_{g,Si}$ = 1.12 eV). This valence electron is shifted into the conduction band and can contribute to the current transport. During the generation process, a hole in the valence band is generated, which can contribute to the current transport as well. If the electron loses its energy, for example, due to electromagnetic waves (photon) or thermal energy (phonon) emission, it drops back to the valence band energy level and recombines with a hole. The generation/recombination effect is the absolutely basic effect which differentiates semiconductor materials from insulator and conductor materials.

To simulate generation/recombination, the SRH model is used:

$$R_{net}^{SRH} = \frac{np - n_i^2}{\tau_n(n+n_1) + \tau_p(p+p_1)}$$

$$n_1 = n_i \exp\left(\frac{E_{trap}}{kT}\right)$$

$$n_1 = n_i \exp\left(\frac{E_{trap}}{kT}\right)$$

where R_{net}^{SRH} is the generation/recombination rate and τ_n and τ_p are the carrier life times of electrons and holes, respectively.

In general, physical model selection for device simulation is an important issue. Depending on the device, one should consider the following for proper selection: For channel length below ~25 nm, some carriers may experience ballistic transport, which increases the driving current by approximately 10%. One option to account for this is to use the 3D Monte Carlo model, which accounts for ballistic carrier transport but requires considerable CPU time. A more practical approach for the analysis and optimization of transistor design, which involves evaluating multiple design trade-offs, is a drift-diffusion model with adjusted carrier velocity saturation. Therefore, the drift-diffusion model is used with velocity saturation:

- Philips unified mobility
- High-field mobility saturation

- Inversion and accumulation layer mobility with auto-orientation

Inversion and accumulation layer mobility is used to model accurate 2D Coulomb scattering in the high channel doping region. A finFET has top and side surfaces where the carriers flow in the on-state. In a typical device orientation, the crystal orientations of the top and side surfaces are different, for example, top surface is (100) and the side surface is (110). The inversion layer mobility has surface orientation dependency. This is taken into account by SDevice using the auto-orientation framework. Inversion and accumulation layer parameters for different surface orientations are well calibrated to experimental data.

2.7.4.1 Thin-layer mobility

When Si channel thickness becomes thinner than approximately 10 nm, low-field mobility is modulated due to the quantum-mechanical confinement effect. Considering the width of the top below 5 nm, the thin-layer mobility model is used.

2.7.4.2 High-k degradation mobility

Since the 45 nm technology node, a high-k metal gate has been used to suppress the gate tunneling leakage current. However, a high-k metal gate degrades the mobility due to

- Remote phonon scattering (RPS)
- Remote Coulomb scattering (RCS)
- Remote dipole scattering (RDS)

The top and side surfaces of finFETs have different crystal orientations. Therefore, the auto-orientation framework is important to consider orientation-dependent quantum correction. Quantum correction parameters are calibrated to the solution of the Poisson–Schrödinger equations by Sentaurus Band Structure.

2.7.5 Stress Effects

Strain engineering has been actively researched and used to enhance transistor performance. The fifth generation of stress-engineering techniques with very high stress levels has been demonstrated in finFETs. Therefore, sophisticated models are required to accurately model stress effects. SDevice provides advanced stress models to take into account the subband modulation by the stress:

- Stress-dependent deformation of band structure
- Strained effective mass and density of states (DOS)
- Subband stress-dependent electron/hole mobility
- Stress-independent carrier saturation velocity

When the carrier velocity is saturated (i.e., when the carrier has sufficient energy to be redistributed), the impact of the stress-induced band structure modulation becomes negligible.

2.7.5.1   Band-to-band tunneling leakage current

The finFET structure tightens the gate control, but it increases the gate-induced drain leakage (GIDL) in the off-state. So, the Schenk band-to-band tunneling model is selected to simulate the leakage mechanism. SiGe models are needed for the SiGe source/drain regions of the p-finFET: Considering that the Ge content in the SiGe source/drain can exceed 50%, one needs to account for the impact of Ge on the source/drain properties.

2.8   Atomistic Simulation

At the nanoscale, the line edge roughness (LER) and random dopant fluctuations are considered as two major sources of short-range variability in aggressively scaled technologies. Predicting the impact of such effects on device and circuit matching performance has been addressed. Gold Standard Simulations (GSS) have announced their process simulator ION. Currently, ION simulates only multidimensional implantation using published data for the projected range and straggle parameters of the relevant impurities in silicon, creating realistic doping profiles. The GSS 3D statistical atomistic device simulator GARAND is used to carry out predictive simulation of statistical variability and reliability. The GSS compact model extractor, MYSTIC, a combination of local optimization and group extraction strategy, is employed to extract a complete nominal set of BSIM4 parameters. The GSS statistical circuit simulation engine RandomSPICE. GSS provides a chain of simulation services that starts with the accurate and predictive simulation of statistical variability and reliability at the transistor level. The main features include 3D simulation of statistical variability, 3D simulation of statistical reliability, statistical compact model (SCM) extraction, and statistical circuit simulation.

2.8.1 GARAND

GARAND is a statistical 3D density gradient-corrected drift-diffusion simulator which can model various sources of statistical variability in contemporary MOSFETs. The main sources of statistical fluctuations are (a) random dopant fluctuations, (b) LER, and (c) metal gate work function variability (WFV). The 3D atomistic device simulator GARAND has been developed over the years by the Device Modelling Group at the University of Glasgow. It has been extensively calibrated against experimental data and other commercial CAD tools. The simulator is based on the drift-diffusion approach to solve self-consistently the Poisson and current continuity equations with density gradient quantum corrections. GARAND has been employed extensively to explore the impact of intrinsic parameter fluctuations on a wide range of devices, spanning from traditional bulk-Si MOSFETs to alternative device architectures, such as SOI and finFETs. The main features of GARAND include:

- Drift-diffusion, Monte Carlo, and nonequilibrium Green's function (NEGF) modules
- The best-available physical models allowing atomic-scale precision, including simultaneous density gradient quantum corrections for electrons and holes
- Mobility models that take into account the discreteness of dopants
- All sources of statistical variability known to be important:
  o Random discrete dopants
  o LER
  o Gate stack granularity
  o Trapped discrete charges
  o Others—custom defined

The variability sources are introduced automatically in the initially idealized device structure through the command line of the GARAND input language. The variability sources available in GARAND include:

- Random discrete dopants in the channel and in the source and drain regions
- LER that can be applied to any geometric object, including the gate, the STI isolation, the fin in the case of a finFET, etc.
- Material granularity, including polysilicon and metal gates
- Interface-trapped charges
- Atomic-scale interface roughness

GARAND has 3D drift-diffusion, Monte Carlo, and quantum transport simulation engines. All simulation engines can simulate identical simulation domains, including, individually or in combination, all statistical variability sources supported by GARAND. The 3D drift-diffusion simulation engine provides accurate results for the statistical variability in the subthreshold region, including threshold voltage and leakage variability. The 3D Monte Carlo simulation engine includes accurate density gradient quantum corrections and ab initio scattering from ionized impurities and interface roughness, essential to simulate accurately the on-current variability. The 3D NEGF quantum transport simulator is essential in sub-10 nm transistors with a current component arising from direct source-to-drain tunneling. It includes also ab initio scattering from ionized impurities and interface roughness, as well as phonon scattering. The visualization modules in GARAND are based on Paraview and include graphing, statistical, and 3D modules.

2.8.2 MYSTIC

Compact models are the bridge between technology and design, and to meet the design challenges of advanced CMOS technologies, they need to accurately capture the statistical behavior of devices. Using automated push button technology the efficient optimization engine and flexible data handling of MYSTIC simplifies the creation of SCM libraries. MYSTIC represents the state-of-the-art in SCM extraction.

The main features of MYSTIC include:

- Advanced optimization engine
- Multistage and multitarget compact model extraction

- Powerful data handling and filtering
- Flexible SCM extraction based on state-machine workflow
- Extraction of nominal models and SCMs
- Automated choice of optimal statistical parameter sets
- Support for BSIM, BSIM-CMG, and PSP
- Seamless integration with GARAND and RandomSPICE

2.8.3 RandomSPICE

The introduction of a reliability-aware design is today accepted as mandatory in order to maintain manufacturing yields. Three main sources of intrinsic parameter fluctuations are considered, namely random dopant fluctuation, gate LER, and metal gate granularity (MGG). Statistics-based predictive analysis of these sources of variability and their adverse impact on the transistor parameters is of great importance from both device and circuit design perspectives. To analyze the impact of individual sources of variability, and their combined effect on device characteristics, all sources should be properly introduced into the device simulator.

RandomSPICE is a statistical circuit simulator. The main features of RandomSPICE include:

- Front end for advanced statistical simulation
- Support for standard SPICE formats, including ngspice, Eldo, and Spectre
- Joint development programs with Mentor and Cadence

2.9 Summary

TCAD has been indicated by the International Technology Roadmap for Semiconductors (ITRS) as one of the enabling methodologies that can support advance of technology progress at the remarkable pace of Moore's law by reducing development cycle times and costs in the semiconductor industry. TCAD approaches yielding accurate physical insight and useful predictive results for real-world semiconductor applications are described using multidimensional (2D and 3D) simulations. In this chapter, a brief background on past and present simulation development activities was given. A brief overview on currently available commercial TCAD tools, viz., Silvaco and Synopsys, has been presented. Nevertheless, the reader should be aware that a number of other comprehensive simulation programs are in existence today. Indeed, several universities have created programs and released them for public use.

References

1. *PCM Studio User Guide*, Version B-2008.06, June 2008.

2. Synopsys Inc., *Paramos User Guide*, Version C-2009.03, March 2009.

3. Synopsys Inc., *Sentaurus Workbench User Guide*, September 2011.

4. Synopsys Inc., *Sentaurus Structure Editor User Guide*, September 2014.

5. Synopsys Inc., *Sentaurus Mesh User Guide*, September 2014.

6. Synopsys Inc., *Sentaurus Device User Guide*, September 2011.

7. Synopsys Inc., *Sentaurus Process User Guide*, September 2011.

8. Synopsys Inc., *Sentaurus Visual User Guide*, September 2014.

9. Synopsys Inc., *Inspect User Guide*, December 2010.

10. Synopsys Inc., *TecPlot SV User Guide*, September 2011.

11. Synopsys Inc., *TSuprem4 User Manuals*, 2006.

12. Synopsys Inc., *Taurus Process User Manuals*, 2006.

13. Silvaco Inc., *UTMOST User's Manual*, 2013.

14. Silvaco Inc., *VWF Interactive Tools, User's Manual*, 2009.

15. Silvaco Inc., *DeckBuild User's Manual*, 2013.

16. Silvaco Inc., *DevEdit User's Manual*, 2013.

17. Silvaco Inc., *ATLAS User's Manual*, 2013.

18. Silvaco Inc., *ATHENA User's Manual*, 2013.

19. Silvaco Inc., *TonyPlot User's Manual*, 2014.

20. Silvaco Inc., *VictoryProcess User's Manual*, 2014.

21. Silvaco Inc., *VictoryDevice User's Manual*, 2014.

22. Silvaco Inc., *VictoryStress User's Manual*, 2014.

23. Silvaco Inc., *VictoryCell User's Manual*, 2014.

24. Smith, C. S. (1954). Piezoresistance effect in germanium and silicon, *Phys. Rev.*, **94**, 42–49.

25. Sverdlov, V. (2011). *Strain-Induced Effects in Advanced MOSFETs* (Springer-Verlag, Wien).

26. Sun, Y., Thompson, S. E., and Nishida, T. (2010). *Strain Effect in Semiconductors: Theory and Device Applications* (Springer Science+Business Media, New York).

27. Bhoj, A. N. (2013). *Device-Circuit Co-Design Approaches for Multi-Gate FET Technologies*, PhD thesis, Princeton University.

Chapter 3

Technology Boosters

Until the late twentieth century, manufacturers sought to reduce unintentional stresses because of their negative effects for devices, for example, by the creation of dislocations promoting leakage current. In recent years, new techniques to improve device performance have emerged based on the stresses induced by the manufacturing process (process-induced stress). In deeply scaled technologies, process and environment variations have become a major concern as stress is a significant source of variability in advanced VLSI technologies that impacts circuit performance. Mechanical stress affects transistor electrical parameter mobility and threshold voltage due to piezoresistivity and stress-induced band deformation, respectively. Unintentional sources of mechanical stress and intentional stress variability cause device performance to depend upon the underlying layout topology and its location in the layout. Consequently, circuit performance becomes highly placement dependent. Thus, it is imperative to capture the effects of layout-dependent stress during circuit analysis. Evaluating circuit performance involves modeling the stress distributions in the layout accurately.

Introducing Technology Computer-Aided Design (TCAD): Fundamentals, Simulations, and Applications
C. K. Maiti
Copyright © 2017 Pan Stanford Publishing Pte. Ltd.
ISBN 978-981-4745-51-2 (Hardcover), 978-1-315-36450-6 (eBook)
www.panstanford.com

3.1 Stress Engineering

3.1.1 Unintentional Mechanical Stress

The unintentional sources of stress can mainly be attributed to the thermal mismatch of the various materials used during integrated circuit (IC) manufacturing. The coefficient of thermal expansion (CTE) of a material determines how fast a material can contract or shrink with decreasing or increasing temperature. Other sources of unintentional stresses that may affect transistor mobilities are caused by wafer/die warpage during wafer processing and thinning, flip-chip package bumps, and CTE mismatch between package substrate and silicon die.

During fabrication of metal-oxide-semiconductor (MOS) devices, unintentional mechanical stresses are generated throughout the manufacturing steps. Moreover, ICs undergo several thermal cycles at elevated temperatures during multiple process steps. During manufacturing, mechanical stresses develop due the thermal mismatch between various layers and constituent materials, thereby causing electrical variations in transistors. Thus, it becomes imperative to consider the contributions of various unintentional stressors on active devices. Mechanical stresses have different origins:

- Oxidation: During oxidation, there is an expansion of the volume of silicon oxide. It induces high local stresses in silicon.
- Ion implantation: Particularly when heavy ions are introduced, local expansions of the crystal lattice occur.
- Annealing: Thermal cycling imposed in furnaces is responsible for deformations caused by the differential thermal expansion coefficients between different materials. These deformations may exceed the elastic limit of materials and thus induce plastic deformations.
- Chemical-mechanical polishing (CMP): CMP can induce local stress, which causes plastic deformation.
- Chip packaging technologies are also responsible for inducing unintentional stress in silicon.

Other major sources of unintentional stress that lie in proximity of the transistors in ICs are:

- Shallow trench isolation, or STI: This is used to isolate transistors in the layout and is the most commonly found source of unintentional stress in the layout. STI is made up of SiO_2 whose CTE differs with that of silicon. STI in the layout surrounds the active transistor regions and can occur in a myriad of shapes, depending upon the neighboring transistors in the layout.

- Through-silicon vias (TSVs) in 3D ICs: TSVs are used to make vertical interconnections between stacked ICs. A TSV is embedded in silicon at an elevated temperature of 250°C and is made up of copper, whose CTE is higher than that of silicon. In the postmanufacturing phase, a thermal residual stress develops in the silicon that is in direct contact with the TSV structure, thus modulating the mobilities of the transistors in near proximity.

3.2   Intentional Mechanical Stress

Technology scaling has exploited the piezoresistive behavior in complementary metal-oxide-semiconductor (CMOS) transistors by deliberately introducing stress in the channels using various techniques. While most of the stress-engineering techniques have been introduced for bulk planar transistors, some of them are scalable to the fin field-effect transistors (finFETs). The sources of intentional stress are summarized as follows:

- Uniaxial source/drain stressors: The source/drain regions of CMOS transistors are recessed and lattice-mismatched alloys are epitaxially grown in the cavities formed. A SiGe alloy with a larger lattice constant than silicon creates beneficial compressive stress along the channel direction for p-type metal-oxide-semiconductor (PMOS) transistors. For n-type metal-oxide-semiconductor (NMOS) transistors, a SiC alloy with a smaller lattice constant than silicon is epitaxially grown in source/drain regions to create a beneficial tensile stress along the channel direction. The source/drain stressors have also been applied for finFETs. This technique is the largest contributor for mobility improvements. The source/ drain stressors in bulk planar transistors and finFETs show dependence on the gate pitch used in the layout, which may vary from technology generations.

- Dual stress liner, or DSL: Dielectric nitride films with intrinsic compressive or tensile stress are grown over the transistor region. While a tensile stress liner is preferred for NMOS transistors, a compressive stress liner is preferred for PMOS transistors. They rely on creating beneficial stress from the vertical direction. However, from the 45 nm technology node onward, their effectiveness was observed to decline. The stress liners have been shown to be not effective for finFETs.

- Stress memorization technique, or SMT: This technique is used for NMOS transistors alone. Here a sacrificial compressive stressed liner is grown on NMOS transistors with polysilicon gate and source/drain regions in the amorphous state. The gate and source/drain regions are crystallized following a rapid thermal annealing step, and the capping stress liner is removed. Even after the stressed capping layer is removed, stress is memorized in the gate and source/drain regions. The gate creates a compressive stress from vertical direction, while tensile stress exists in the source/drain regions. The SMT has also been demonstrated for finFETs.

- Source/drain contact stress: Tensile stress can similarly incorporated in the metal contacts over source/drain regions of an NMOS transistor. The metal contacts are deposited by creating trenches in source/drain regions. However, the effectiveness of this technique is diminishing in sub-45 nm technologies, which use raised source/drain regions to reduce source/drain resistance.

- Replacement metal gate and gate-last process: This method has been shown to be effective for bulk planar transistors and finFETs. In advanced technologies, metal gates are employed instead of polysilicon gates to improve threshold voltage control. First a sacrificial polysilicon gate is deposited, and subsequent fabrication steps for source/drain epitaxy and salicidation are completed. Then the polysilicon gate is stripped off, thereby increasing the stress transferred into the channels. Subsequently, the metal gate is deposited in the gate terminal region. Using certain process conditions the metal gate can be incorporated with tensile or compressive strain, which acts vertically on the channels. A metal gate with compressive stress is preferred for NMOS transistors, while a metal gate with tensile stress is preferred for PMOS transistors.

3.3   Stress-Engineered Transistors

Strain technology has been successfully integrated into CMOS fabrication to improve transistor performance, but the stress is nonuniformly distributed in the channel, leading to systematic performance variations. A high-mobility channel, providing a high drive current, and heterostructure confinement, providing good short channel control, make an interesting combination for future technology nodes.

3.3.1   CESL

In this section, we focus on the process of inducing mechanical stress in MOSFETs using the contact etch stop layer (CESL), which does not require any additional step in the standard manufacturing process (Fig. 3.1). We shall examine how a nitride layer deforms the silicon in the conduction channel. The purpose of the CESL is to change the crystal lattice of silicon. To understand the transmission mechanism of the CESL in the channel, three areas may be distinguished: the nitride which is located at the top of the gate will be the area's *top CESL*, the part on the spacers on the sidewalls of the gate is the *lateral CESL*, and the layer that covers the source and drain will be the *CESL bottom*. These different areas are shown in Fig. 3.2. If the nitride film is in tension, the bottom layer just shoots the source and drain, which will cause a compression of the silicon at the base of this layer. The top layer will transmit stress throughout the gate, which will generate a compressive stress in the plane of the center of the channel. The lateral CESL is just in the plane (X and Y) of the channel and its major influence will be perpendicular to the channel. The deformations in the plane of channel are therefore due to the portion of the layer which takes on the CESL source and drain and that the part of the top which, via the gate, induces a compressive stress in the central part of the channel (Fig. 3.3).

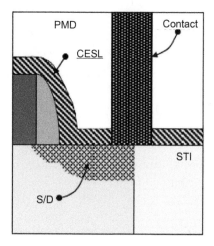

Figure 3.1   Schematic presentation of the contact etch stop layer (CESL).

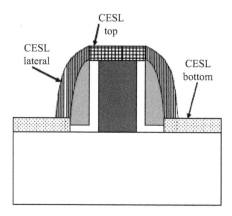

Figure 3.2   Diagram representing the three different regions of the contact etch stop layer (CESL).

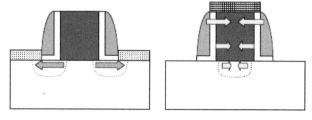

Figure 3.3   Representation of the stress effects of the CESL bottom (left) and CESL top and bottom combined (right) of the nitride film on the transistor channel.

The second effect of the etch stop layer (ESL) stress is in the direction perpendicular to the plane of the channel. Typical stress fields generated by a tensile CESL are depicted in Fig. 3.4 for long transistors. The CESL induces mainly a compressive stress whatever the stress components are. Figure 3.4 shows the stress in the silicon under the gate electrode in the Z direction when the CESL is in tension. In this case the channel of the transistor is in compression. Thus, considering the CESL film in three parts (Fig. 3.2), the strain transmission takes place directly and through the side of portion CESL. The state of stress along the Z direction is quite uniform and compressive, whereas along the X and Y directions, there are two distinct areas, namely (1) a compressive one near the channel center and (2) a tensile one near the channel edges. A compressive stress is mostly localized near the spacers. Due to the elasticity of the silicon, the stress in the Z direction results in a channel deformation in three directions. The compressive area takes up about two-thirds of the channel and the tensile area takes up about one-third. Contrary to long transistors, the CESL induces a tensile stress along the X and Y directions, whereas stress in the Z direction remains compressive.

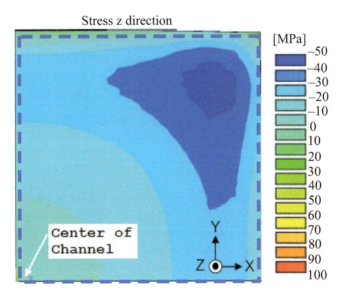

Figure 3.4 Stress in the channel (represents ¼ of the gate region, W = 1 μm, L = 1 μm) in the Z direction, perpendicular to the plane of the channel [5].

The nonuniformity of stress field in the X and Y directions for long transistors as the uniformity for short transistors can be explained by considering a corner effect. The corner effect is due to the interaction of the lateral CESL and the bottom CESL. It is responsible for the stress field disruptions observed near the bottom edge of spacers. Figure 3.5 presents the stress evolution from a long device to a short one. It shows that by decreasing the gate length, the size of the compressive area (in the X and Y directions) is reduced, whereas the size of the tensile area remains the same. For a short device, this corner effect has a major impact, which explains the state of stress in the X and Y directions. For long devices, the corner effect is responsible for nonuniformity of the stress field.

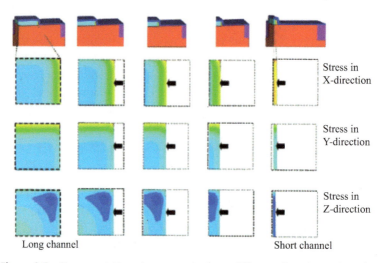

Figure 3.5   Representative stress map in three different directions when the gate length is decreased [5].

The different impacts of these three zones can be exploited to reach the best stress configuration that will enable one to have the best performance for both NMOS and PMOS devices. In general, a compressive CESL is needed to optimize PMOS, and a tensile CESL is needed to optimize NMOS. As such, the performances of NMOS and PMOS transistors cannot be enhanced using the same CESL. Also, the stress map in the channel is affected by the layout. Thus for managing the stress configuration in the channel by an engineered CESL, one needs to control and optimize by adjusting either the thickness or/and the intrinsic stress of the different CESL zones. Thus, the suitable combination of these three zones leads not only to the maximum electrical performance for NMOS and PMOS but also to a wide range of layouts.

3.3.2 STI Stress

In the following section, we discuss the consequences of stress due to STI on various types of MOS transistors. We consider mainly the effects of dimensions, different types of substrate orientation, and new materials for STI. The simplest way to check the impact of the stress of STI is to consider transistors all having the same electrical characteristics and to vary the intensity of the strain in the conduction channel by changing the distance between the gate and the trench, while keeping a constant distance between the contacts and the gates (the electric field lines then remain unchanged). Figure 3.6 shows the schematic representation of the stress effect. When the length of the source and drain regions reduces, compressive stress is generated due to STI, which propagates toward the channel and modifies the carrier conduction parameters. The change in length of the source and drain zones will give rise to deviations of threshold voltage due to a change in the channel current. For NMOS transistors, when the distance between the gate and the trench isolation decreases, an increase in compressive stress occurs in the X direction, leading to a current degradation. It has been shown that the variation of the mobility as a function of stress applied in any direction and, in this case, with a compressive stress in the direction of the current (<110> here), only hole mobility enhances. For compressive stress in the X direction (that of the gate length), it is beneficial for the mobility of holes in the <110> direction.

3.4 Hybrid Orientation Technology

Hybrid orientation technology (HOT) is a method which involves using two different crystallographic planes for two types of devices. Plane (110) is used for the PMOS transistors and (100) plane for the NMOS. The starting substrate is an silicon-on-insulator (SOI) substrate which has (a) a substrate part with a given orientation (e.g., 110) and (b) a layer of SOI with a different orientation (e.g., 100). Figure 3.7 shows the main processing sequence for fabrication. Using this method one can obtain the best orientation for each of the devices. However, the problem is that one obtains a p-type device (or n-type) on SOI, while the other n-type (or p-type) is on a solid substrate. This can be problematic for designers who need to use both types of architectures in MOS for the same circuit.

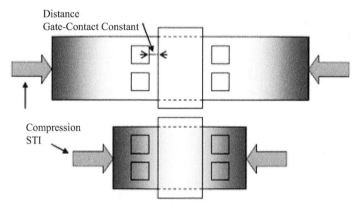

Figure 3.6 Schematic representation of the effect of stress when STI reduces the distance between the gate and the trench [6].

3.5 High-*k*/Metal Gate

As scaling has reached fundamental material limits, especially for gate oxides, further scaling can be realized only by introducing new materials with a high dielectric (high *k*) constant. The main motivation for the migration to high-*k* materials is to continue scaling the equivalent oxide thickness of devices, while maintaining a low leakage current. It is clear that high-*k* materials are promising for microelectronic devices. There are a number of materials with a *k* value much higher than that of SiO_2. Aluminum oxide (Al_2O_3) is one high-*k* candidate having overall superior behavior for MOSFET applications, except its comparatively low *k* value. High-*k* materials have a wide array of applications other than as gate oxides. Some of their important applications are listed below:

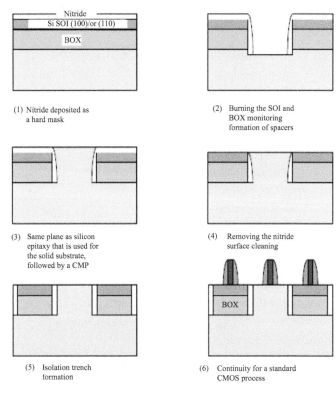

Figure 3.7 Description of hybrid orientation technology (HOT) [6].

- Gate dielectrics
- High-aspect-ratio diffusion barriers for Cu interconnects
- Adhesion layers
- Highly conformal coatings for microfluidic and microelectromechanical systems (MEMS) applications
- Coating of nanoporous structures
- Other nanotechnology and nanoelectronic applications
- Fuel cells, for example, single-metal coatings for catalyst layers
- Bio-MEMS
- Electroluminescence
- Protective coatings
- Storage capacitor dielectrics

- Pinhole-free passivation layers for organic light-emitting diodes (OLEDs)

There is a set of material and electrical requirements for a viable alternate high-k gate dielectric material. Major requirements include:

- Larger energy bandgap with a higher barrier height to the Si substrate and metal gate to reduce the leakage current
- Large dielectric constant value
- Good thermodynamic stability on Si to prevent the formation of a low-k SiO_2 interface
- Good kinetic stability
- High amorphous-to-crystalline transition temperature to maintain a stable morphology after heat treatment
- Low oxygen diffusion coefficients to control the formation of a thick, low-k interface layer
- Low defect densities in high-k bulk films and at the high-k/Si interface with negligible C–V hysteresis (<30 mV)
- Low fixed charge density ($\sim 10^{10}$ cm^{-2} eV^{-1})
- Low high-k/Si interface state density ($\sim 10^{10}$ cm^{-2} eV^{-1})
- High-enough channel carrier mobility ($\sim 90\%$ of SiO_2/Si system)
- Good reliability and a long lifetime

A conventional CMOS process flow consists of a mature set of subsequent process steps. In the initial steps of processing, the NMOS and PMOS areas are separated by local oxidation of silicon or more advanced STI and doped by ion implantation. The SiO_2 gate dielectric is formed by thermal oxidation of Si, and poly-Si is then deposited atop. The poly-Si is subsequently patterned and etched to form the gate electrode. Source and drain areas are implanted to form extension junctions in a self-aligned process where the poly-Si gate electrode is simultaneously implanted. After reoxidation of the gate area, spacers are formed and subsequent deep implantations are performed, followed by thermal activation of implanted dopants inside both source/drain and gate areas by means of rapid thermal processing (RTP). The salicidation process is then performed to form silicide in the source, drain, and gate. The final steps in the front-end processing are capping of the structure and formation of contact holes.

The introduction of high-k and alternative gates has been, is, and will continue to be challenging. New processes and integration concepts will be required. First of all, a single high-k material (and its corresponding gate electrode material, if not poly-Si) and its deposition technique should be identified. Among various deposition techniques for high-k gate dielectrics, ALD appears to be the most suitable and viable deposition technique. For two types of gate electrodes, ALD TiN and in situ doped p⁺poly-SiGe, process steps have been developed at the Royal Institute of Technology, Sweden. The process flow for fabrication of p-MOSFETs with high-k materials and a TiN gate is shown in Fig. 3.8 and the process flow for p-MOSFETs with high-k materials and a p⁺ poly-SiGe gate in Fig. 3.9.

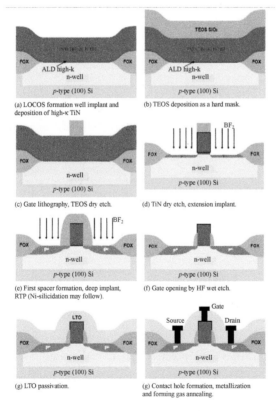

Figure 3.8 Process flow for p-MOSFETs with a TiN/high-k gate stack [7].

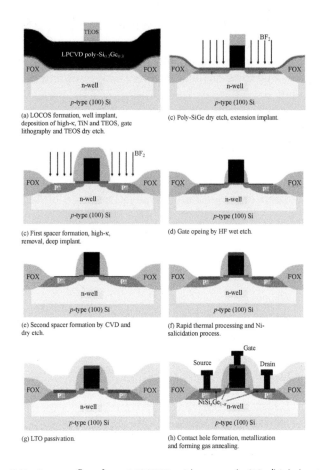

Figure 3.9 Process flow for p-MOSFETs with a p+ poly-SiGe/high-*k* gate stack [7].

3.6 Stress Evolution during Semiconductor Fabrication

Most of the semiconductor manufacturing process steps are done at high temperature with deposited thin films. Maximum temperatures are usually higher in the front end of line (FEOL) and can reach, at least in a short time pulse, about 1000°C. As for the backend-of-line (BEOL) operations, temperatures are kept under 400°C to save the global thermal budget. Since the coefficients of thermal expansion of the films are different, the cooling back to ambient temperature induces stress within the layers and the substrate. The force balance leads to both partial stress relaxation of the thin film, and wafer curvature. It is important to understand the stress transfer mechanisms, and predictive simulation is necessary to calibrate models from device layout to enhanced mobility. Producing an electronic device is based on a large number of steps, while chips are batch-processed on undiced silicon wafers. FEOL consists of stages that allow making the core devices and achieve electrical functionalities into the silicon. After contacting the transistors together, comes BEOL, which handles the interconnections. Beyond the manufacturing processes themselves, device integrity must be ensured. Using technology computer-aided design (TCAD) the evolution of the stress state during the whole process flow can be studied.

As the race for miniaturization is now reaching its limitations, to stick with the performance specification challenges, alternative critical dimension (geometric) downscaling has become essential. Stress engineering is a low-cost way to boost CMOS technology. Mobility can be changed depending on the state of stress in the channel [1]. Strain engineering has its own limitations as the stress generated cannot be infinite and may lead to various detrimental effects such as defect nucleation, dislocation generations, or microstructural changes, which would lead to performance degradation, such as leakage, opens, or shorts, and finally affect device integrity.

Since the 90 nm CMOS technology node, the strained nitride capping CESL is used as a stress-engineering booster enabling transistor improvement. The effectiveness of the nitride layers constrained with the source/drain SiGe, SOI substrates, and the effect of channel direction will be discussed for MOSFET gate lengths up to 14 nm. In this chapter, we present a simulation study explaining how the CESL transmits its intrinsic stress to the Si channel. It is demonstrated that the CESL stress transmission is the outcome of several CESL parts acting separately (direct effect) or in association (indirect effect). Finally, some guidelines are given for an optimization of CESL use.

3.6.1 Stress Modeling Methodology

Several options have been proposed in the literature to account for the modifications in mobility under stress [2, 3]. The first one is the so-called piezoresistive approach, which relates the strained to the unstrained mobility. It is used to relate mobility to stress as it provides information on the lattice parameters, which are related directly to transport properties. This approach has the disadvantage as it is applied to the overall carrier mobility, leading to various sets of piezocoefficients, depending on the doping and electric fields. The second alternative relates the change in mobility to the effective conduction mass, omitting thereby the impact of the reduced intervalley scatterings. In this study, the CESL intrinsic stress is tensile and equal to 1 GPa.

It is assumed to be a biaxial stress following the curvature of the deposited layer. This biaxial stress has been implemented in the model as an intrinsic stress oriented along a local work plane, whose orientation follows the CESL curvature. To analyze the relative contribution of the strained CESL only, all the materials are stress free except the nitride capping layer. Thus, the stress state of Si channel is here only due to the CESL intrinsic stress. An elastic behavior is assumed for all materials so that the contribution of other stress sources like STI could simply be added to the CESL one. The CESL has been split in three distinct zones making up of the whole nitride capping layer:

- Top CESL above the gate region
- Lateral CESL above the spacer region
- Bottom CESL above source/drain and STI regions

Thus, the thickness and intrinsic stress of each zone can be different from others. The classical CESL is the sum of these three CESL zones having the same thickness and the same intrinsic stress. The aim of this virtual splitting is to determinate the contribution of each CESL zones on the channel state of stress along the X, Y, and Z directions.

Typical stress fields generated by a tensile CESL are depicted in Fig. 3.10. The CESL induces mainly a compressive stress whatever the stress components.

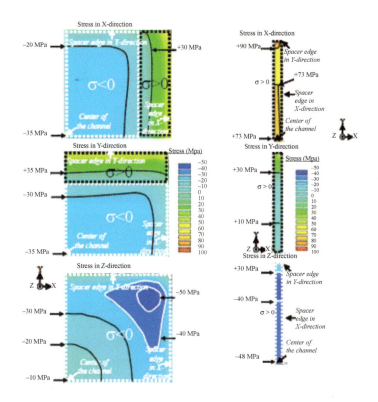

Figure 3.10 Top view of a quarter of the channel stress fields for long devices [5].

3.6.2 Stress Evolution during Thick Stress Layer Deposition

Any semiconductor device manufacturing process consists of multiple steps, including material layer depositions (in some cases with specific intrinsic stress), etching of the layers or their portions, as well as heating and cooling cycles. Thus, in reality the stresses in the device structure change from step to step because of geometry, materials, and temperature, which are involved throughout the process sequence. Accurate stress simulation has become necessary for semiconductor device performance and reliability analysis.

Commonly one calculates stresses in the final device structure with different stressor regions by specifying values of intrinsic stresses, which is known as the one-step model, where stresses are calculated only once.

To accurately predict stresses in thick layers the process should be considered as a series of deposition and relaxation steps to emulate mechanical quasi-equilibrium during this physical deposition process. The more detailed step-by-step model calculates stresses after each process step, taking into account current geometry, temperature, and material properties, as well as stresses generated in the structure during previous steps. This approach is known as the stress evolution or stress history model. Obviously, the stress history approach is more time consuming but it definitely provides more accurate results. The following example shows how simulation can help to boost electrical performance. To investigate the CESL stress transmission, finite element simulations are commonly performed using Ansys software.

In the following section, we use VictoryProcess and VictoryStress to demonstrate how the stress evolution model can be applied to simulation of stresses generated during the thick layer deposition process and compare the results with the standard one-step model. The deposition of thick layers can be considered as a series of deposition and relaxation steps to emulate mechanical quasi-equilibrium during the physical deposition process.

The simulation consists of five steps:

1. Formation of a test structure with a one-layer nitride stressor in ATHENA
2. Calculation of stresses in the test structure using the one-step model of VictoryStress
3. Formation of a test structure with a multilayer stressor in ATHENA
4. Calculation of stresses using the stress evolution model of VictoryStress and the looping capability of DeckBuild
5. 2D visualization and average stress extraction for both models

The stress evolution simulation uses 20 layers at each step, while the effect of the new sublayer deposition is emulated by changing the status of this new sublayer from nonactive with properties inherited from air to the active material region with properties inherited from nitride and the intrinsic stress of 1 GPa. Then VictoryStress recalculates stresses in the modified structure after each deposition step. The simulation results are shown in two plots which compare S_{xx} and S_{yy} stresses for two models. In flat areas of the deposited layer both models produced similar and essentially uniform stress. In the single-layer case a nonuniform stress region appears only near the corners of the spacer/nitride. However, in the case of multilayer stress history simulation a highly nonuniform stress field extends diagonally from the spacer/substrate corner to the top of the film. In other words, stresses are significantly higher where the layers are sharply bent. As the result the integrated stresses under the gate appear to be 1.4–2 times higher when the stress history method is used. The extracted averaged stresses are calculated and plotted in Fig. 3.11.

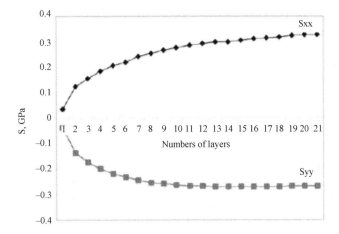

Figure 3.11 Dependence of S_{xx} and S_{yy} stresses (averaged under the gate) on the number of 8 nm sublayers. Lateral stress becomes more compressive and vertical stress becomes more tensile, both asymptotically approaching certain limits.

Channel stress simulations using single- and multilayer models were performed for various nitride CESL films with intrinsic stress ranging from highly compressive to highly tensile. Piezoresistivity constants for different wafer orientations were used for mobility enhancement evaluation [4] in VictoryStress simulation. Mobility enhancement factors are calculated (Table 3.1) by use of average stress values and piezoresistivity factors from the following relations:

$$\frac{\Delta\mu_{xx}^{TG}}{\mu} = \frac{2}{3}\frac{\Delta\mu_{xx}^{DG}}{\mu} + \frac{1}{3}[(1+\pi_{11}\sigma_{xx})\times(1+\pi_{12}\sigma_{yy})\times(1+\pi_{13}\sigma_{zz})-1]$$

$$\frac{\Delta\mu_{xx}^{DG}}{\mu} = (1+\pi_{11}\sigma_{xx})\times(1+\pi_{12}\sigma_{yy})\times(1+\pi_{13}\sigma_{zz})-1$$

Table 3.1 Mobility enhancement factors calculated by use of average stress values and piezoresistivity factors

Material	n-type		p-type	
10^{-12} **dynes/cm²**	**<100>**	**<110>**	**<100>**	**<110>**
π_{11}	−102.2	−31.1	6.6	71.8
π_{12}	53.4	−17.5	−1.1	−66.3
π_{13}	53.4	53.4	−1.1	−1.1

Figure 3.12 shows contours of lateral stress in a simulated structure after deposition of a nitride film with compressive intrinsic stress, simulated using single- and multilayer models. In flat areas, both simulations give identical stress values in the nitride. However, stress distributions are very different near and around the gate. In the single-layer case, the conformal free surface is smooth and has fewer corners. Thus, the stress in the nitride is relatively uniform with one noted region of stress concentration located at the surface of the nitride associated with the spacer/substrate corner. On the contrary, stress is much uniform in the film simulated using the multistep approach. The inner layers of the nitride film more closely follow sharp contours of the gate structure and the overall CESL shape. Stress significantly increases in the areas where the layers sharply bend. This region of high compression extends diagonally from the film top to the spacer bottom. As a consequence stresses in the spacer and the gate become very nonuniform as well, changing from tensile on the top to compressive at the bottom, and

the magnitude of compressive stress in silicon underneath the gate increases. Lateral stress becomes more compressive and vertical stress becomes more tensile, both asymptotically approaching certain limits.

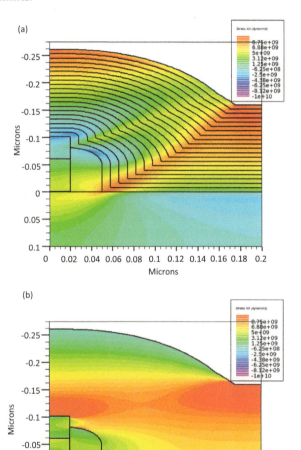

Figure 3.12 The S_{xx} distribution for stress evolution model contours of lateral stress (Pa) in a transistor structure using (a) 20-layer deposition of 160 nm compressive intrinsic stress and (b) single layer.

3.6.3 Stress Evolution in Thick Stress Layer Deposition in 3D

The next simulation is essentially a 3D version of the earlier 2D simulation. The stress evolution simulation uses the same structure with 20 layers at each step. The test structure used in this example is a quarter of a 40 nm MOS transistor. The X direction is the gate length direction, the Y direction corresponds to the width of the channel, and the Z direction represents the vertical direction.

The input deck consists of five sections:

- Formation of a test structure with a one-layer nitride stressor in VictoryProcess. The structure is shown in the first 3D plot.
- Calculation of stresses in the test structure using the one-step model of VictoryStress.
- Formation of a test structure with a multilayer stressor in VictoryStress. The structure is shown in the second 3D plot.
- Calculation of stresses using the stress evolution model of VictoryStress and the looping capability of DeckBuild.
- 3D and 2D visualization of simulation results for both models.

The material parameters used in the one-step model simulation are:

Material nitride intrin.sig = 1.0×10^{10}
Material silicon young.m = 1.67×10^{12}
Material silicon poiss.r = 0.28
Material poly young.m = 1.67×10^{12}
Material poly poiss.r = 0.28
Material oxide young.m = 1.0×10^{12}
Material oxide poiss.r = 0.25
Material aluminum young.m = 0.715×10^{12}
Material aluminum poiss.r = 0.35

The simulation results are shown in two 2D plots which compare S_{xx} stresses for two models (Fig. 3.13). The S_{xx} stresses are also compared in two cut planes: YZ (vertical plane which roughly corresponds to the 2D simulation shown in Fig. 3.14) and XY (horizontal plane just 1 nm under the silicon surface). The first of the 2D comparison plots is very similar to the 2D simulation, as shown in Fig. 3.13. It shows that the stress evolution model predicts a higher stress level under the gate and is better illustrated in the second XY plane. The upper-left corner of the figure is the cut plane corresponding to the gate area. The simulation results are shown in two 3D plots which compare S_{xx} stresses for two models, as shown in Figs. 3.15 and 3.16.

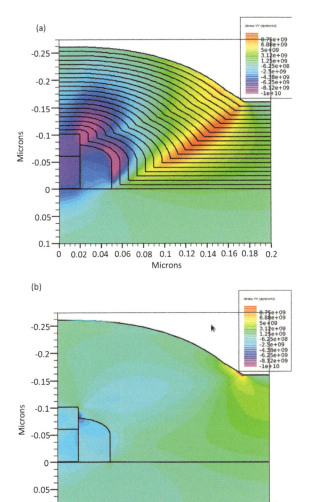

Figure 3.13 Contours of lateral stress (Pa) in 40 nm transistor structures with a 20 nm source/drain recess simulated using (a) single-layer and (b) 160-layer deposition of nitride with compressive intrinsic stress. The S_{yy} distribution for stress evolution model: (a) 20-layer deposition of 160 nm compressive intrinsic stress and (b) single-layer deposition.

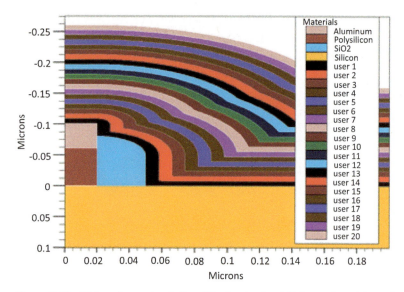

Figure 3.14 VictoryStress simulation (2D view) using 160-layer deposition of nitride with compressive intrinsic stress.

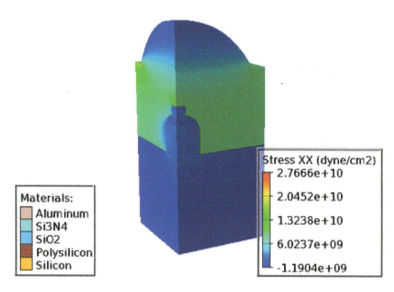

Figure 3.15 VictoryStress simulation using single-layer deposition with compressive intrinsic stress.

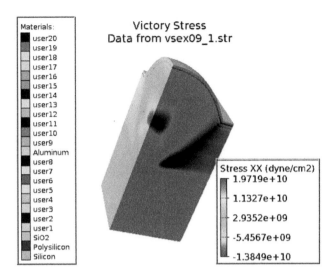

Figure 3.16 VictoryStress simulation using 160-layer deposition of nitride with compressive intrinsic stress.

3.7 Summary

Strain engineering continues to evolve and will remain one of the key performance enablers for future generations of CMOS technologies. In this chapter, we have discussed the challenges involved in the heterogeneous integration of novel strain-engineering techniques that have remarkably advanced the silicon CMOS transistor architecture, including embedded SiGe (e-SiGe), embedded SiC (e-SiC), SMT, dual stress liners (DSLs), and stress proximity technique (SPT). We have presented a simulation procedure explaining how the CESL transmits its intrinsic stress to the Si channel. It has been shown that the CESL stress transmission is the outcome of several CESL parts acting separately (direct effect) or in association (indirect effect), including the corner effects for small transistors. The stress transfer mechanisms for long- and short-channel transistors are explained. Finally, some guidelines are given for optimization of the usage of the CESL. Channel stress simulations using single- and multilayer models were performed for various nitride ESL films with intrinsic stress ranging from highly compressive to highly tensile. It is observed that the stress history model used for stress simulation in a thickly deposited stressor film gives more accurate results.

References

1. Maiti, C. K., and Maiti, T. K. (2012). *Strain-Engineered MOSFETs* (CRC Press, Taylor and Francis, USA). Shin, K. (2006). *Technologies for Enhancing Multi-Gate Si MOSFET Performance*, PhD thesis, University of California, Berkeley.

2. Sun, Y., Thompson, S. E., and Nishida, T. (2010). *Strain Effect in Semiconductors: Theory and Device Applications* (Springer Science+Business Media, New York).

3. Ortolland, C. (2006). *Etude des effets des contraintes mécaniques induites par les procédés de fabrication sur le comportement électrique des transistors CMOS des noeuds technologiques 65nm et en deça*, PhD thesis, L'Institut National des Sciences Appliquées de Lyon.

4. Sverdlov, V. (2011). *Strain-Induced Effects in Advanced MOSFETs* (Springer-Verlag, Wien).

5. Sun, G. (2007). *Strain Effects on Hole Mobility of Silicon and Germanium p-Type Metal-Oxide-Semiconductor Field-Effect-Transistors*, PhD thesis, University of Florida.

6. Fiori, V. (2010). *How Do Mechanics and Thermomechanics Affect Microelectronic Products: Some Residual Stress and Strain Effects, Investigations and Industrial Management*, PhD thesis, L'Institut National des Sciences Appliquées de Lyon.

7. Ortolland, C. (2006). *Etude des effets des contraintes mécaniques induites par les procédés de fabrication sur le comportement électrique des transistors CMOS des noeuds technologiques 65nm et en deça*, PhD thesis, L'Institut National des Sciences Appliquées de Lyon.

8. Wu, D. (2004). *Novel Concepts for Advanced CMOS: Materials, Process and Device Architecture*, PhD thesis, Royal Institute of Technology.

Chapter 4

BiCMOS Process Simulations

Semiconductor technology is a broad and complex field encompassing many activities. Modern processes may have roughly 300 steps that are carried out on every product wafer. Semiconductor processes are continually modified to improve yield and to improve device characteristics. The primary objective of process simulation is to accurately predict the physical/structural layers and geometry of devices at the end of a process run, as well as the active dopant/stress distributions. As shown in Fig. 4.1, the input to process simulation is a process flow guided by process assumptions and layout/layer masks. The initial wafer/substrate is subject to a variety of process conditions, each of which may involve steps like oxidation, diffusion, implantation, deposition, and etching. Lithography simulation is also performed to accurately capture feature geometries. Process simulation generally uses a finite-element or finite-volume mesh to compute and store the device dopant and stress profiles. Every geometric change in the simulation domain requires a new mesh that fits the new device boundaries in order to model the next series of process steps.

Process simulation is especially helpful in the initial phase of technology development. As device lots become more and more expensive, process modeling is increasingly important. Process simulation and modeling is increasingly sophisticated, but accuracy remains a problem. There is generally a time lag between the

Introducing Technology Computer-Aided Design (TCAD): Fundamentals, Simulations, and Applications
C. K. Maiti
Copyright © 2017 Pan Stanford Publishing Pte. Ltd.
ISBN 978-981-4745-51-2 (Hardcover), 978-1-315-36450-6 (eBook)
www.panstanford.com

introduction of a particular process and its accurate modeling. Modeling of front-end processes has been facing challenges caused by a continuous reduction of the implant energies and of the thermal budgets of the annealing schemes from soak anneals to the spike anneals which are now used for production. The complexity of physical models is a major factor that impacts process simulation. Simplified physics minimizes computation time. With technology scaling, however, the need for ever more accurate doping/stress profiles has increased and complex physical models are added at each new generation. Also, the best models available are usually too complicated and slow to be used in multidimensional process simulation, so often compromises have to be made. It has to provide insight during design, optimization guidelines during implementation to manufacturing, and debug during large-scale manufacturing.

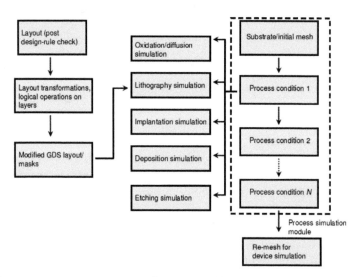

Figure 4.1   Generic process simulation steps [1].

The main objective of a device simulation is to understand the behavior of a device with different characteristics and under various bias conditions. Main elements of device simulation are device structure, material system parameters, circuit/contact boundary conditions, a list of physical effects to be captured, numerical constraints on the solver, a carrier transport model, and the modes of simulation. In device simulation, first we need to describe the geometric structure of the device. Then we discretize the geometry region. In a device simulator a semiconductor device is modeled through a geometric structure with certain physical properties which are governed by a set of partial differential equations. Finally we solve the system of equations by applying some numerical schemes, such as the finite-difference method or the finite-element method. To fully understand the electrical response of a semiconductor device, different device simulations are performed. A device simulator is used to perform DC, AC, and transient mixed-mode simulations. DC simulations provide I_d–V_g and I_d–V_d characteristics, while AC analysis generates C–V characteristics. The resulting I–V plots are analyzed to extract electrical parameters. Single-device simulation is used to investigate transport phenomena in a single device. Mixed-mode simulation is used to study the behavior of small circuits constructed out of individual device instances and is generally less rigorous in terms of physical models, owing to the increase in simulation complexity. In the following sections, we perform simulations of some basic unit processes such as ion implantation, oxidation, deposition, etching, and photolithography by using standard models.

4.1   Ion Implantation Simulation

Simulation of ion implantation is the most commonly used method for doping. This technique stood up to other doping techniques such as epitaxy and spin-on-dopant sources. However, doping by ion implantation causes several problems, for example, creating defects in the crystal lattice and amorphization of silicon. Ion implantation may be simulated using analytical models based on empirical point-response distribution. In the case of 2D simulations, an ion beam incident at a point is assumed to generate a distribution function. The final ion concentration at a certain point is calculated by computing the superposition of all distribution functions of all possible points of incidence on the surface of the implanted structure. While the vertical distribution function is only dependent on the depth, the lateral distribution depends on both, the lateral spread and the depth which determines the shape of the doped region. Two major vertical distributions of the implanted ions are used in simulators for calculating the primary doping profile:

Gaussian distribution:

$$f_p(x) = \frac{1}{\sqrt{2\pi}\sigma_p} \exp\left(-\frac{(x-R_p)^2}{2\sigma_p^2}\right),$$

where the projected range R_p is the first moment and σ is the standard deviation of the projected ranges. The skewness of the profile is zero in this approximation, while the fourth moment, the kurtosis of the Gaussian profile, is 3.

Pearson distribution, which satisfies the relation

$$\frac{d}{dx}f_p(x) = \frac{x-R_p-a}{b_0+b_1(x-R_p)+b_2(x-R_p)^2}f_p(x),$$

where b_0, b_1, b_2, and a depend on the second and the third moment. In simulators, mainly type IV, type V, type VI, and dual Pearson distributions are used for calculating the final doping profile. To account for the correlation between the lateral and the vertical ion scattering, the standard deviation σ_2 of the lateral ion spread is, in general, assumed to be dependent on the depth:

$$f_1(y,x) = \frac{1}{\sqrt{2\pi}\sigma_1(x)} \exp\left(-\frac{y^2}{2\sigma_1^2(x)}\right)$$

To minimize channeling of the ions through the silicon lattice, a tilt of 7° is used in the ion implantation setup of planar devices. Arsenic ions are used to form the n+ regions, while boron ions were used to form the p$^+$ regions. Due to very thin body thicknesses used for the silicon-on-insulator (SOI) transistors, very low implantation energies in the range of 1 keV to 5 keV for arsenic and 0.2 keV to 1 keV for boron are commonly used in the simulation of ion implantation.

Simulation of activation and diffusion of implanted dopants is an important issue, as the amount of the active doping concentration has a big impact on the device performance. Generally, diffusion and activation of implanted dopants is done by annealing at high temperatures above 800°C. The implanted dopants diffuse by means of either interstitials or vacancies, dependent on the kind of dopant. Additionally, they are built in the silicon lattice, where they can contribute to the current transport, that is, they are activated. Furthermore, annealing at high temperatures leads to recrystallization of amorphous layers and to elimination of point defects resulting from ion implantation. To calculate the diffusion of the implanted dopants, the diffusion current may be computed using the well-known Fick's first law with the field-dependent diffusion. The implanted dopants diffuse by means of point defects of the silicon lattice, either interstitial or vacancies. Therefore, the effective diffusion coefficient of any respective dopant-point-defect pair has to be calculated by the Arrhenius equation:

$$D_{N^c} = D_0 \cdot \exp\left(-\frac{E_a}{kT}\right)$$

where D_0 is the Arrhenius prefactor of the specific dopant point defect pairs and E_a is the respective activation energy. The redistribution of the implanted dopants can be expressed by an extension of Fick's second law of diffusion. For modern complementary metal-oxide-semiconductor (CMOS) devices, rapid thermal annealing (RTA), or millisecond annealing (MSA) techniques are commonly used. Two examples of such RTA and MSA schemes are the so-called spike annealing and flash annealing. Depending on the requirements of different device architectures, the most feasible annealing scheme has to be chosen. To simulate diffusion and activation of arsenic and boron by using RTA or MSA, calibrated simulation models have to be used. From the point of view of process simulation, transient phenomena in diffusion and activation, especially of boron, were for long the main issue. Although these effects are qualitatively understood, their quantitative modeling is still a problem.

In ion implantation, ions collide elastically with target atoms, creating ion deflections, energy loss, and displaced target atoms (recoils). Channeling is caused by ions traveling with few collisions and little drag along certain crystal directions. Ions come to rest after losing all the energy on elastic collisions (nuclear stopping) and inelastic drag (electronic stopping). Modeling of ion implantation needs to include the following:

- Ion energy loss mechanisms
- Ion range distribution
- Ion channeling in crystalline silicon
- Implantation-induced damage modeling

Figure 4.2 shows a comparison of Gaussian, Pearson, and secondary ion mass spectrometry (SIMS)-verified dual Pearson SVDP methods. This example compares implant analytical models: Gaussian (symmetrical) profile, single Pearson (amorphous implant), and SVDP method. In Fig. 4.3, a comparison of zero-tilt boron implant profiles obtained using the SVDP method with experimental results is shown.

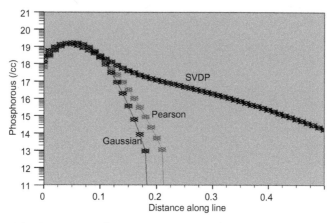

Figure 4.2   Comparison of Gaussian, Pearson, and SVDP methods.

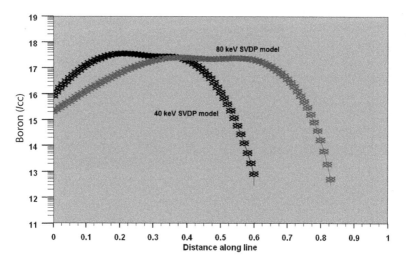

Figure 4.3   Experimental verification of channeling profiles.

Oxidation-enhanced diffusion of boron is studied next (Fig. 4.4). The incorporation of models that account for the codiffusion of point defects (interstitials and vacancies) with dopant impurities addressed the important physical effects of oxidation-enhanced diffusion and transient-enhanced diffusion, forming the basis for the impurity diffusion models used in modern silicon process simulation programs. Such programs combine many different physical modeling capabilities, including oxidation, diffusion of various dopants and their interactions, etching, deposition, and various approaches to simulating ion implantation.

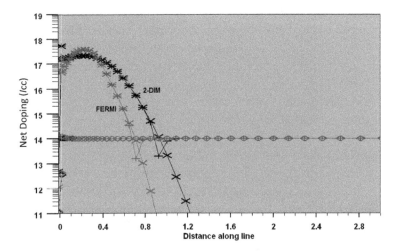

Figure 4.4   Oxidation-enhanced diffusion of boron.

4.2   Optical Lithography Simulation

Lithography is a critical pattern transfer process step for the fabrication of small device structures, and it is vital to understand the limitations and effects of process variations and design parameters. The cost of lithography equipment is very high, and as the process development time for new lithography methods is large, the cost-effectiveness of computer modeling of lithography processes is very important. We introduce the concepts which are essential to understanding how the optical lithography process is simulated.

The lithography imaging system consists of a wafer upon which the structures are patterned, a photomask (reticle) which contains the desired pattern, and a wafer stepper, the optical equipment which transfers the pattern from the mask to the wafer. The mask contains transparent and opaque areas on a glass mask substrate with which the light from the source is patterned. The wafer is coated with an optically sensitive polymer photoresist. In a positive photoresist, the actinic light interacts with the resist, breaking polymer bonds and increasing its solubility in a developing solution. In a negative resist, the incident light causes bonds to form, thus decreasing the local solubility rate. Resists are then baked to diffuse somewhat the soluble and insoluble areas. The wafer stepper contains the machinery and optics for aligning the wafer and mask, exposing the mask, and imaging the resulting optical pattern onto the photoresist-covered wafer.

The lithography system can be divided into (a) a light source, (b) the propagation of light through free space, (c) the patterning of the light by the mask, (d) the imaging of the patterned light by the projection lens, and (e) the interaction of the light with the materials on the wafer. The underlying optics of the lithography system can be simplified for simulation purposes, as shown in Fig. 4.5. Lithography modeling involves development of lithographic model incorporating particularly parameters for the light source and 3D mask effects. One needs to develop mask synthesis algorithms that are robust against mask variations and incorporate regularization methods in solving inverse problems in imaging to tackle mask complexity. Technology computer-aided design (TCAD) may be used to investigate quantitatively relevant types of variability: systematic and random variabilities that arise from nonidealities of the lithographic process. Three-dimensional simulations are performed in conjunction with the modeling of the lithography process. In this section, we perform a simple lithography simulation. This example demonstrates the first step in lithography simulation. A simple mask (Figs. 4.6, 4.7, and 4.8) has been used: two elbows and a contact hole (CD = 1 μm). The stepper used in the simulation was the model Canon FRA-1550 with the following parameters: g-line, numerical aperture (NA) = 0.43, and $\sigma = 0.5$.

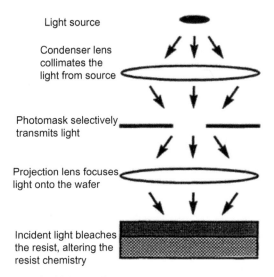

Figure 4.5 Simplified lithography system.

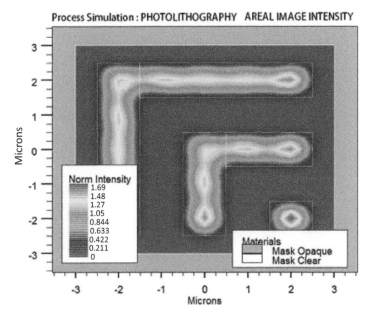

Figure 4.6 Exposure system: numerical aperture (NA) = 0.43; partially coherent g-line illumination (wavelength = 435 nm). No aberrations or defocusing is found. The minimum feature size is 1 μm.

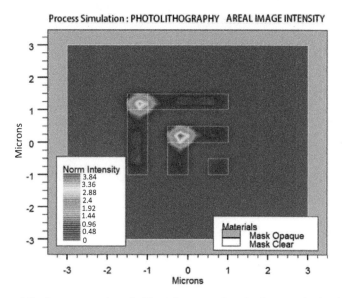

Figure 4.7 Same example as in Fig. 4.6. except that the feature size has been reduced to 0.5 µm. Note the poorer image.

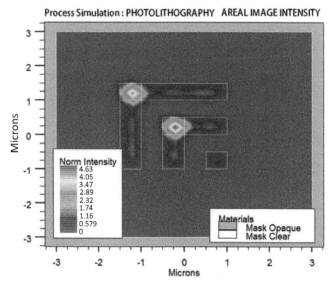

Figure 4.8 Same example (Fig. 4.6), except that the illumination wavelength has now been changed to i-line (wavelength = 365 nm) and the NA has been increased to 0.5. Note the improved image.

4.3   Contact-Printing Simulation

The simulation of contact printing for a simple mask consisting of a contact hole (CD = 1 μm) is performed next. The parameter *gap* in the image statement specifies the mask-to-wafer gap (in microns) for contact printing. Figure 4.9 shows the areal image intensity.

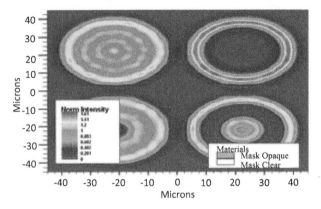

Figure 4.9   Contact-printing simulation.

4.3.1   Nonplanar Lithography

In this example we show nonplanar lithography simulation. The effects of exposing a photoresist over a change in topography are demonstrated. The example uses a user-defined aerial image cross section. Figures 4.10 and 4.11 show the important stages during simulation. The photoresist is developed and effects of the light reflected off the 45° slope can be seen.

4.4   BJT Process Simulation

In this section, the different fabrication technologies of bipolar junction transistors (BJTs) are explained with emphasis on the double poly-Si technology. For bipolar transistors, a variety of different process flows are used by manufacturers, unlike for CMOS technology, where every company seems to follow a unified process flow. The main bipolar process steps are:

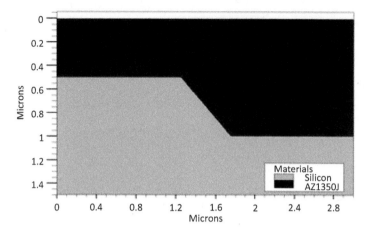

Figure 4.10 Nonplanar lithography simulation (photoresist development).

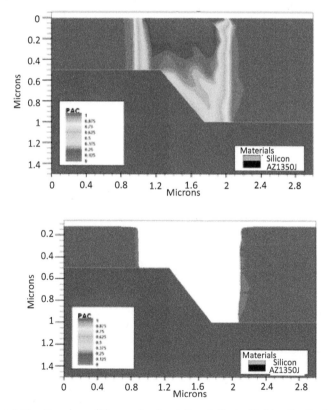

Figure 4.11 Nonplanar lithography simulation (effect of light reflection).

1. Isolation trench definition
2. Subcollector implantation
3. Subcollector activation
4. Poly-Si base–contact deposition
5. Active area base opening
6. Base implantation
7. Collector implantation
8. Nitride spacer definition
9. Metallization
10. Final anneal

4.4.1   Polysilicon Emitter Bipolar Technology

In the single-polysilicon process, the main advantage is the possibility of direct integration in a CMOS process. The bipolar complementary metal-oxide semiconductor (BiCMOS) process has the advantages of having both bipolar, n– and p–metal-oxide-semiconductor field-effect transistors (MOSFETs), in the same process. A simplified process flow is shown in Fig. 4.12. The double-poly-Si bipolar technology uses not only an n+*doped poly-Si emitter but also an extrinsic base to be contacted with a p+-doped poly-Si layer. This feature enables full self-alignment of the active layers of the transistor. A schematic process flow for a double-poly-Si technology is shown in Fig. 4.13. As can be seen in Fig. 4.13a, the first lithography step of the process is the local-oxidation-of-silicon (LOCOS) mask, which defines the active areas and the collector contacts. The second mask is the collector plug implantation mask (Fig. 4.13b). After deposition and boron implantation of the base poly-Si, an oxide is deposited, which serves as an isolating layer between the emitter and the base poly-Si. Then the base poly-Si mask is exposed and both the oxide layer and the base poly-Si layer are etched. Figure 4.13c shows the structure after removal of the resist, followed by a drive-in anneal for forming the extrinsic base regions. The intrinsic base is then implanted through the emitter window. After spacer formation the emitter poly-Si is deposited and implanted. The emitter poly-Si also serves as a contact to the collector window. The emitter poly-Si and the underlying oxide are then etched after exposure and development of the last mask. Finally, the uncovered areas of both the base and the emitter poly-Si are silicided.

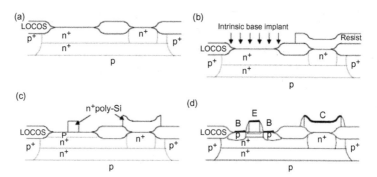

Figure 4.12 Schematic process flow for a single-poly-Si bipolar transistor [2].

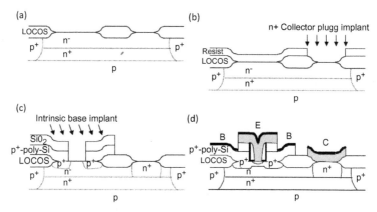

Figure 4.13 Schematic process flow for a double-poly-Si bipolar transistor [2].

In the following bipolar process simulation example, we use ATHENA for process simulation to obtain the bipolar device structure, remesh the structure in DEVEDIT, and use ATLAS simulation to obtain the electrical characteristics of a poly-Si emitter NPN bipolar transistor with two base contacts. As shown in Fig. 4.13, the typical process flow for an NPN bipolar transistor is provided in ATHENA. The final device structure obtained is shown in Fig. 4.14. The first boron implant forms the intrinsic base region (Fig. 4.13c). The structure has a heavy n^+ emitter, a 1.0×10^{18}/cc peak base concentration, a buried collector layer, and a heavy p^+ extrinsic base contact (Fig. 4.15). As the mesh used for process simulation is not optimal for use with device simulation, to recreate a mesh that has zero obtuse triangles in the semiconductor region, the mesh generation tool DEVEDIT is used to refine the mesh. The remeshed final device structure used in the ATLAS simulation is shown in Fig. 4.16. In the ATLAS simulation bipolar models used includes concentration-dependent mobility, field-dependent mobility, bandgap narrowing, concentration-dependent lifetimes, and Auger recombination.

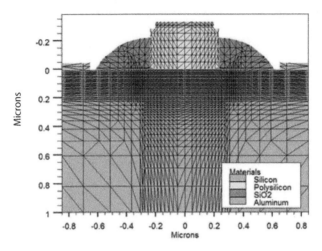

Figure 4.14 ATHENA (process)-simulated BJT device structure.

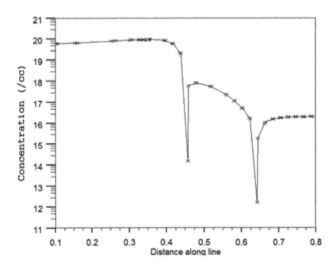

Figure 4.15 Doping profile in a BJT.

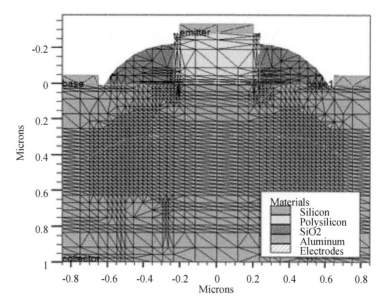

Figure 4.16 DEVEDIT-gridded BJT structure.

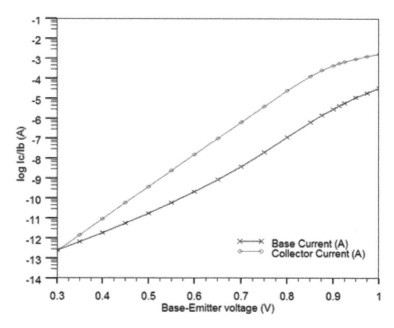

Figure 4.17 ATLAS-generated Gummel plot of a BJT.

Electrical parameters are then extracted from the Gummel plot (Fig. 4.17) and the DC output characteristics (I_C vs. V_{CE}) for different constant values of base currents, while V_{CE} is ramped (Fig. 4.18). The Gummel plot is obtained by applying a bias ramp on the base electrode up to 1.0 V. The parameter AC on the solve statement sets the AC analysis on. The frequency of this signal is set to 1 MHz.

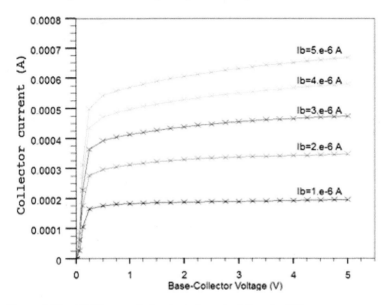

Figure 4.18   ATLAS-generated output characteristics of a BJT.

4.5   3D MOS Process Simulation

For decades, TCAD has been limited to 2D due to a lack of computing power for the simulator, and earlier device structures had little variations in the third dimension. However, nowadays 3D simulation is becoming increasingly important due to pronounced 3D effects present in state-of-the-art devices which are truly 3D in structure. The need for 3D simulation tools has become indispensable in the past few years when scaling limits of bulk CMOS technology have boosted research on alternative, essentially 3D architectures, for example, multiple-gate devices (MuGFETs). Three-dimensional TCAD is now made possible because of advances in computer hardware.

Three-dimensional TCAD gives us an opportunity in device design and new insights into their behavior. We are able to view the devices as they actually exist rather than using 2D cross sections. In this section, we discuss the methodology used to generate 3D structures from process simulation using VictoryProcess. VictoryProcess is a general-purpose layout-driven 1D, 2D, and 3D process simulator including etching and deposition, implantation, diffusion, and oxidation simulation capabilities.

To perform a TCAD simulation, we first need the mask layout for all the process steps. The mask layout can be designed in a variety of drafting tools like Autodesk, AutoCAD, and specialized EDA tools like Cadence Virtuoso and Tanner L-Edit. These tools are used to create a GDSII file (.gds), which is the current industry standard for integrated circuit (IC) layout work. GDSII files are imported from a layout by the MaskViews program. MaskViews is an IC layout editor designed to interface the IC layout with other Silvaco tools. It can draw and edit an IC layout, store and load complete IC layouts, and import/export layout information using the industry-standard GDSII and CIF layout format. MaskViews provides layout information to the simulators, enabling any part of a layout to be simulated. Currently, supported simulators are the SSUPREM3 1D process simulator and the ATHENA 2D process simulator. MaskViews provides a set of mask regions for each layout level, giving the start and end points of masks on any arbitrary cross section on the layout. MaskViews also provides information on how ATHENA should construct its grid.

MaskViews can be used as a viewer of GDSII-formatted files, besides the basic function of being an interface between layout and process simulator. Various settings are available, including the choice of which layers to import into a new design. The scale of the pattern being imported is critical: the user can choose to convert the units used in the GDSII file and can also specify a scaling factor, if required. The basic purpose of MaskViews is to create files that will be used in the subsequent 3D process simulation. When GDSII files are not available, MaskViews can also be used to create the geometric shapes and layers used in the process simulation.

In a VictoryCell simulation the process etching steps are driven from a layout file generated from MaskViews (see Fig. 4.19). This file will be included in the support files when this example is loaded. VictoryCell is used in the following example to simulate a full 3D process and device flow for a 45 nm n-MOSFET. The process simulator takes the input files created by MaskViews and performs 3D process simulation based on the user's command input. After finishing the simulation, the process simulator can export its output to the device simulator. The device simulator will then use the exported mesh and material information to perform electrical, thermal, and other simulations. The output of the device simulator can be viewed directly with a plotting GUI or saved to a graphic format such as postscript for convenient batch processing.

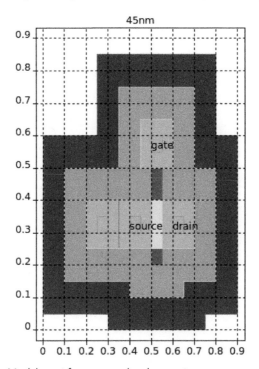

Figure 4.19 Mask layout for process development.

The polysilicon gate, spacer, and electrodes of the final MOSFET structure simulated using VictoryCell and MaskViews is shown in Fig. 4.20. The cross section along the source–drain electrodes showing net doping is shown in Fig. 4.21.

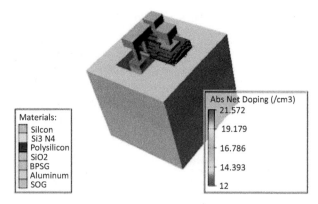

Figure 4.20 Polysilicon gate, spacer, and electrodes of the final MOSFET structure simulated using VictoryCell and MaskViews.

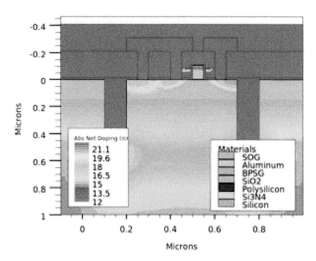

Figure 4.21 Cross section along source–drain electrodes showing net doping.

4.5.1 VictoryProcess

VictoryProcess integrates four core modules:

- Etching and deposition
- Oxidation and stress
- Implantation
- Diffusion

Different crystal planes of silicon are known to have different oxidation rates, for example, the silicon plane with Miller indices <111> is oxidized approximately 1.7 times faster than the <100> plane. During an oxidation process the geometry of a 3D structure may change significantly and the silicon/oxide interface may pass through various crystal planes with different oxidation rates. Therefore, the orientation and type of the silicon wafer affect the resulting geometry of an oxidized structure on this wafer. VictoryProcess can be used to model the anisotropic behavior of silicon during oxidation. VictoryProcess offers several modes for the simulation/emulation of an oxidation process, namely

- The analytical oxidation mode
- The empirical oxidation mode
- The full physical oxidation mode

The oxidation module includes:

- Simulation in empirical, full physical, or hybrid mode
- Empirical mode applied for very thin oxidation layers
- Deal-Grove and Massoud models used in empirical mode
- Simulation in full physical mode of oxidant transport, reaction on the Si/SiO_2 interface, viscous flow, material deformation, and stress formation
- Automatic switching between empirical and full physical modes, depending on oxide thickness
- Empirical mode used in planar regions with coarse mesh, allowing layer thicknesses smaller than mesh size to be resolved
- Full physical mode used in regions with fine mesh
- Stress-dependent oxygen transport and interface reaction
- Orientation dependence, doping dependence, and ambient conditions

VictoryProcess includes a module for plasma etching. The module is designed to simulate plasma etching processes at the feature-scale size. The simulation in the reactor-scale region is out of the scope of VictoryProcess. All transport characteristics data (as functions of reactor parameters needed for the feature-scale simulation) are modeled by user-definable C functions and are supplied to the module. The plasma etching simulator shares many elements with the standard physical etching/deposition module, such as:

- The topology of a given feature is defined by its layers of various materials, as described by level set functions given on Cartesian meshes
- All fine details of the structure are captured on embedded finer meshes, and automatic and/or manual adaptive mesh refinement is available.
- Particles fluxes, etch rates for different materials, and types of particles involved in the process are modeled by appropriate functions, which are implemented in the user-accessible C-Interpreter model library.
- All feature topological changes caused by the plasma etching process during a given time are captured by the solution of the corresponding partial differential equations acting on the level set functions.
- A feature's structure is automatically updated after the simulation time has expired.

The physical etching module includes:

- Selective etching of several materials
- Comprehensive set of default models
- Selective isotropic etching
- Selective anisotropic etching
- Selective directional etching
- Plasma etching
- Ion milling
- Material-dependent yield functions
- Redeposition
- Rotating beams and static beams
- Reactive ion etching
- Deep reactive ion etching
- Ballistic transport of reactants
- High performance due to multithreading

Physical deposition includes the following models:

- Comprehensive set of default models
- Conformal deposition

- Nonconformal deposition
- Directional deposition
- Selective deposition
- No deposition on selected materials
- Ion beam deposition
- User-accessible yield functions
- Rotating beams and static beams
- Ballistic transport of reactants

The oxidation step is decomposed as a succession of the following basic processes:

- Diffusion of oxygen through the oxide
- Solving the diffusion equation
- Reaction at the SiO_2/Si interface
- Volume expansion
- Interface propagation
- Deformation of the structure according to the mechanical behavior of each material
- Viscous flow
- Solving the creep flow problem

The diffusion module features include:

- Fermi diffusion model compatible with ATHENA/SSuprem4
- Fick diffusion model for nonsemiconductor materials
- Simulation of multiple dopant diffusion
- Accounts for solid solubility, dopant activation, and segregation
- Simulation of transient enhanced diffusion effects
- Three-stream and five-stream diffusion models
- Point defect trapping and clustering models
- Impurity segregation at all material interfaces
- Impurity activation and solid solubility
- Simulation of oxidation-mediated diffusion

The ion implantation module features include:

- Analytical and Monte Carlo method for ion implantation
- Analytical method using a wide range of implantation conditions

- Gaussian profiles, Pearson profiles, and dual Pearson profiles
- User-definable profiles
- By means of moments of the profile
- By means of data files
- Range-scaling techniques to take into account multimaterial layers
- High-performance numerical algorithms and multithreading

The Monte Carlo method takes into account the following important implantation effects:

- Ion channeling
- Ion dose dependency due to damage accumulation
- Effect of multiple layers of different materials
- Partial shadowing of ion flux for tilted implants

The set of default models includes:

- Direct, Fermi, five-stream, single-pair
- Dopant activation and solid solubility
- Impurity segregation at material interfaces
- Point defect trapping
- Point defect clustering

VictoryProcess is used for the investigation of a 3D process and device as it is suitable for many semiconductor technologies such as fin-shaped field-effect transistors (finFETs). There are several aspects of basic unit processes such as implantation, oxidation, etching, and deposition process characteristics which can be obtained by applying the fast operational modes of VictoryProcess. Among these characteristics are:

- Rounded corners
- Tapered sidewall
- Nonconformal epitaxial growth
- Selective geometrical etch
- Selective chemical-mechanical polishing (CMP)

Three-dimensional process simulation for an NMOS transistor with shallow trench isolation is considered. It may be noted that the parameter *mirror* does create one half of the device with source and drain. Process simulation is performed for a quarter of the device. The model used in simulation is the *fullcpl* model for diffusion simulation, and point defects are taken into account during annealing. The fully coupled oxidation diffusion mode is used. The simulation starts in 2D and switches over to 3D later in the flow. Enhanced diffusion due to implantation-induced damage as well as oxidation-enhanced diffusion is taken into account as the *fullcpl* model is used. The main process steps for MOSFET fabrication are:

1. Initial substrate
2. Gate oxide deposition
3. Threshold voltage adjustment implant and gate poly
4. Lightly doped drain (LDD) implant and nitride spacer
5. Source/drain implant
6. Mirror structure and export to the device simulator
7. Contact definitions for device simulation

Specifications of the device are as follows:

- $L_g = 0.1$ (gate length)
- $W_g = 0.2$ (gate width)
- $X_w = 0.35$ (x direction analysis region)
- $Y_w = 0.4$ (y direction analysis region)
- $Z_w = 2.0$ (substrate depth)
- Dep = 0.3 (trench depth)
- $H_{poly} = 0.1$ (poly height)
- $Z_{top} = 0.3$ (analysis region top)
- $t_{Si3N4} = 0.05$ (Si_3N_4 sidewall thickness)

The simulation procedure includes:

1. Defining of simulation domain and wafer material
2. Simulation mesh resolution and substrate doping
3. Defining of mask layers for the process simulation
4. Trench etching
5. Etching and deposition with geometrical mode
6. Saving of simulation status and export to TonyPlot format for visualization

Figures 4.22 through 4.34 show the VictoryProcess-generated structures for the main process steps described above for a MOSFET.

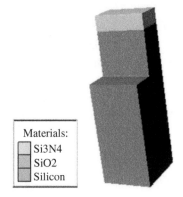

Figure 4.22 Oxidation of trench sidewall and annealing in dry oxidation ambient.

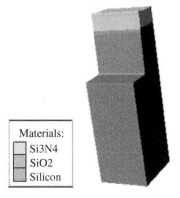

Figure 4.23 Fill in trench/CMP and etching, deposition, and CMP with geometrical mode.

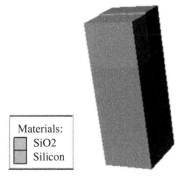

Figure 4.24 Channel implantation through scattering oxide and deposition with geometrical mode and implantation with analytical mode.

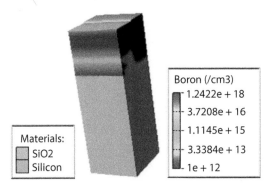

Figure 4.25 Removal of scattering oxide and annealing in wet oxidation ambient.

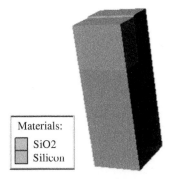

Figure 4.26 Switch to 3D simulation, gate patterning, and deposition with geometrical mode and mask.

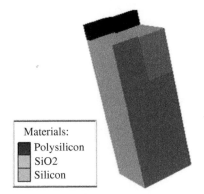

Figure 4.27 Poly reoxidation, annealing in dry oxidation ambient, halo implantation, implantation with Monte Carlo mode, extension implantation, implantation with Monte Carlo mode, and activation of the cluster damage.

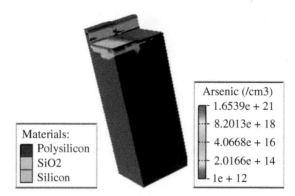

Figure 4.28 Gate sidewall, and etching and deposition with geometrical mode.

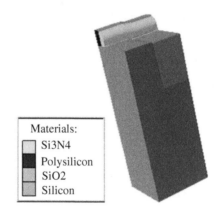

Figure 4.29 N-S/D implantation and implantation with analytical mode.

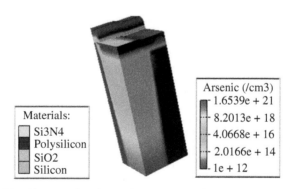

Figure 4.30 S/D anneal and annealing in inert ambient.

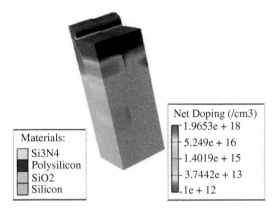

Figure 4.31 End front-end process flow, export structure for device simulation, and add mask layer for contact formation.

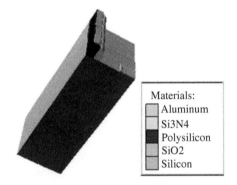

Figure 4.32 Export simulation status to a format and a mesh suitable for VictoryDevice.

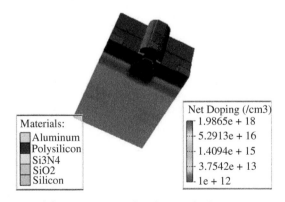

Figure 4.33 Final device structure showing net doping.

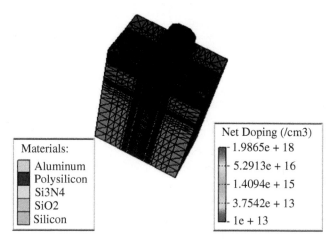

Materials:

- Aluminum
- Polysilicon
- Si3N4
- SiO2
- Silicon

Net Doping (/cm3)
- 1.9865e + 18
- 5.2913e + 16
- 1.4094e + 15
- 3.7542e + 13
- 1e + 13

Figure 4.34 Final device structure showing the grid generated by VictoryProcess.

4.6 Summary

A CMOS is the most widely used discrete device structure in the semiconductor sector. Process optimization for manufacturability is one of the most challenging issues in the semiconductor industry since the adoption of the submicron CMOS technology. The CMOS process was optimized for understanding the overall process and technology limitations. In this chapter, both bipolar and CMOS technology issues, such as threshold voltage and saturation current control, and methods to optimize the process through a series of characterization studies were discussed.

References

1. Bhoj, A. N. (2013). *Device-Circuit Co-Design Approaches for Multi-Gate FET Technologies*, PhD thesis, Princeton University.

2. Linder, M. (2001). *DC Parameter Extraction and Modeling of Bipolar Transistors*, PhD dissertation, Kungliga Tekniska Högskolan, Stockholm, Sweden.

Chapter 5

SiGe and SiGeC HBTs

The International Technology Roadmap for Semiconductors (ITRS) predicts required target performances of SiGe/SiGeC HBTs for addressing the future need of emerging applications. Figure 5.1 shows the cutoff frequency, f_T, and the maximum oscillation frequency, f_{max}, the ITRS targets for SiGe HBTs, together with an area indicating the best experimental frequency performance achieved so far. The ITRS longer-term roadmap calls for HBTs operating at f_T/f_{max} of 460/500 GHz in 2015 and 570/610 GHz in 2020. Bandgap engineering concepts that were previously only possible to implement in compound semiconductor technologies, have now become viable in silicon technology (see Ref. [8] of Chapter 1). The property of $Si_{1-x}Ge_x$ that is of interest for bipolar transistors is the bandgap, which is smaller than that of Si and controllable by varying the Ge content. State-of-the-art circuits involving SiGe HBTs are primarily in data communication and radar systems at 24, 60, and 77 GHz where SiGe HBTs are operated close to $f_T/3$.

Currently many research groups are moving SiGe HBTs into the cutoff frequency range close to 500 GHz and above, enabling the future development of communication, imaging, or radar-integrated circuits working at frequencies up to 160 GHz. Currently, these bands are dominated by III–V technologies like GaAs HEMTs. The concept of bandgap engineering in Si technology provides a new degree of freedom in the design of the base, allowing the base doping to

Introducing Technology Computer-Aided Design (TCAD): Fundamentals, Simulations, and Applications
C. K. Maiti
Copyright © 2017 Pan Stanford Publishing Pte. Ltd.
ISBN 978-981-4745-51-2 (Hardcover), 978-1-315-36450-6 (eBook)
www.panstanford.com

be increased and the base width to be reduced, while at the same time maintaining a reasonable value of gain. In this way, much higher values of f_T and f_{max} may be achieved. The main drawbacks in existing SiGe HBT designs are the necessary high bias currents leading to high power dissipation and limited noise. To achieve low noise there is a need for transistors with higher f_T and f_{max}.

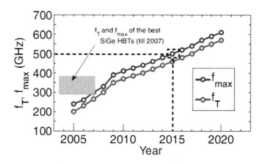

Figure 5.1 ITRS f_T and f_{max} targets for SiGe HBTs, together with the region showing the best SiGe HBTs available [1].

As silicon and germanium are completely miscible over the entire compositional range, bandgap engineering can be profitably used in silicon technology. The lattice constant difference of Si and Ge is about 4.17%, and if one is grown on the other below the critical thickness, the layer is strained. This strain has been used to vary the bandgap energy, band discontinuities, and effective mass to split the valley degeneracy and other electronic properties. For the last several years, silicon-germanium has shown great promise for novel device applications. In the near future, SiGe devices and Si/SiGe superlattice-based devices may even play an important role in the integration of complex electronic circuitry with optoelectronic functionality on a single integrated circuit (IC) chip. SiGe is helping to meet the challenges of the mobile communication market by providing high integration, high performance, low noise, low current consumption, and outstanding efficiency.

The addition of substitutional carbon to silicon-germanium thin films leads to a new class of semiconducting materials (SiGeC). This new material can remove some of the constraints (such as the critical layer thickness) on strained $Si_{1-x}Ge_x$ and may open up a new field of device applications for heteroepitaxial Si-based systems.

The incorporation of carbon can be used to enhance the SiGe layer properties, to obtain layers with new properties, and to control dopant diffusion. The incorporation of carbon allows one to use a higher boron dose within the SiGe base layer and/or narrower undoped SiGe spacers, leading to significantly improved transistor performance.

The introduction of a low concentration of carbon ($<10^{20}$ cm^{-3}) in the SiGe region of SiGe HBTs has several significant effects, a compensation of the compressive strain in the lattice, a reduction of the boron out-diffusion from the base region, and a change in the bandgap being the most important to HBT devices. The presence of carbon also relaxes technological process design constraints by reducing the sensitivity of dopant profiles to subsequent processing steps. When compared to SiGe technologies, the addition of carbon offers significantly greater flexibility in process design and a greater latitude in processing margins.

HBTs outperform BJTs in their high-frequency performance, owing to the following reasons. Firstly, for the HBT, the emitter being doped lighter than the base leads to significant lowering of the emitter junction capacitance, which, in turn, reduces the emitter charging time. Another important degree of freedom available in HBTs is in increasing the base doping. The facility to increase base doping to almost the solubility limit of the material enables significant reduction in base sheet resistance. This is quite unlike the BJTs, where emitter doping must exceed that of the base to have a practically useful current gain. The performance advantages of SiGe HBTs over Si BJTs for RF/microwave applications are mainly the high cutoff frequencies. Further, by incorporating the grading of the energy gap in the base, HBTs can develop large quasi-electric fields in the base at higher base-doping levels. This would significantly minimize base transit time, thereby boosting cutoff frequency values.

When optimized, all-silicon devices show a typical cutoff frequency of 50 to 70 GHz, while the corresponding SiGe device switches at approximately two to three times that speed. In the SiGe HBTs the presence of germanium in the base region reduces the bandgap of the base. The reduction in the base bandgap of a SiGe HBT lowers the potential barrier to electron injection into the base and thus exponentially increases the number of electrons injected from emitter to base for a fixed bias, even if the base is doped more heavily than the emitter. Although part of this high gain is sacrificed to a very high base doping concentration necessary to achieve a low base resistance compared to a Si BJT, it offers an additional degree of freedom, which relaxes a series of trade-offs affecting the device design. Several key advantages of SiGe HBTs over conventional bipolar devices include a higher cutoff frequency, improved noise performance with a higher cutoff frequency, and high early voltage and gain product.

Two types of SiGe HBTs have been demonstrated, a graded-base HBT and a box-profile HBT. In both processes, the addition of Ge in the base allows for a reduced base transit time for a given base sheet resistance, thus providing a device with a simultaneously high cutoff frequency and maximum oscillation frequency. A higher base doping concentration also provides advantages in both a higher early voltage (due to less modulation of the space region into the neutral base) and low noise due to a low intrinsic base resistance, which translates into performance advantages for RF applications. The effect of SiGe on the base bandgap and transport properties can be advantageously used in the bipolar transistor to provide a higher collector current and reduce the carrier transit time. As a SiGeC alloy has a lower bandgap than Si, a bipolar transistor could be created with SiGeC in the base to obtain much higher values of current gain.

The cross section of a representative SiGe HBT is shown in Fig. 5.2. Although the inherent band offset caused by the Ge profile occurs in the valence band, it is effectively translated to the conduction band. With a constant p-type doping in the base, both the Fermi level and the energy difference between the Fermi level and valence band are fixed; the Ge grading induces a valence band offset, but because the Fermi level must remain constant in equilibrium, it must decrease in energy along with the conduction band edge. For DC operation, one fundamental impact of the graded conduction band offset is to enhance minority electron transport across the base by inducing a drift field. In addition, the Ge content at the emitter–base (EB) junction will reduce the potential barrier for electron injection from the emitter to the base, yielding exponentially greater electron injection for the same applied V_{BE} (i.e., higher current gain). Finally, a finite Ge content at the collector–base (CB) junction will positively influence the output conductance of the transistor (i.e., higher early voltage), since the smaller base bandgap near the CB junction effectively weights the base profile so that back-side depletion of the neutral base with increasing V_{CB} is suppressed. For AC operation, the Ge grading-induced drift field will intuitively lead to a reduced base transit time, which typically is the limiting transit time that determines performance metrics such as the maximum operating frequency. In addition, the Ge-enhanced injection of electrons from the emitter into the base dynamically produces a back-injection of holes from the base into the emitter. This reduces the emitter charge storage delay time, which is reciprocally related to the AC current gain of the transistor. To understand the differences between the SiGeC HBT and the Si BJT the band diagrams of both are compared in Fig. 5.3.

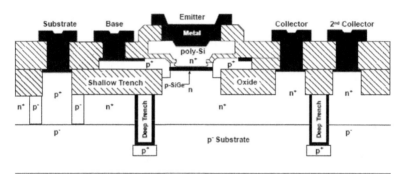

Figure 5.2   Cross section of a representative first-generation SiGe HBT [2].

These DC and AC effects are dependent on the profile of the Ge content, especially the mole fraction at the EB junction and the degree of grading across the neutral base. However, trade-offs in profile design exist because of the fact that SiGe film stability limits the total Ge content that can be present. Consequently, different Ge profiles can be designed to achieve specific performance goals. For example, a triangular profile beginning at the EB junction and peaking just inside the CB space charge region would maximize the frequency performance and early voltage, while providing little improvement to the current gain. A box-shaped profile that is flat across the base would maximize the DC current gain but would not enhance electron transport across the base. Alternately, a trapezoidal profile 3 would simultaneously improve all performance metrics, albeit to a lesser extent. At present, state-of-the-art n-p-n SiGe HBTs have been demonstrated with peak f_T and f_{max} above 400 GHz at room temperature.

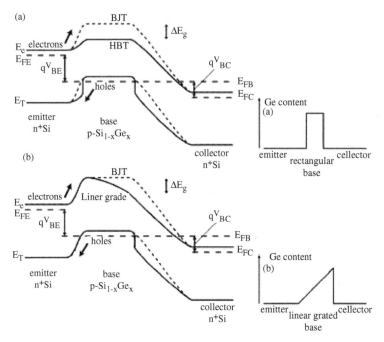

Figure 5.3 Comparison of the band diagrams of a SiGeC HBT (solid line) and a Si bipolar transistor (dashed line). (a) The band structure for a Si bipolar transistor and a rectangular Ge base profile (HBT). (b) The band structure for a Si bipolar transistor and a linearly Ge-graded SiGe HBT [3].

After the introduction to the merits of SiGe HBTs and a brief look at the status of SiGe world, we consider in this chapter the following:

- The development of a TCAD-based comprehensive simulation framework for developing a scaling roadmap for SiGeC
- The understanding and modeling of linearity in SiGeC devices and circuits
- The development of a TCAD-based methodology for reliability stress analysis and predictive virtual aging of SiGeC devices and circuits

5.1 SiGe HBTs: Process and Device Simulation

In this section, we consider the fabrication and analysis of a SiGe heterojunction bipolar transistor (HBT) using the ATHENA-ATLAS combination. A bipolar device with a SiGe base is created in ATHENA and then passed to ATLAS for electrical analysis to extract a Gummel plot and other device characteristics. The simulated device structure can be directly used as input for device simulation; however, a regridding procedure is generally used for robust device simulation. Using DeckBuild's autointerface the process simulation structure has been passed into ATLAS automatically. This autointerface therefore allows global optimization from process simulation to device simulation to SPICE model parameter extraction. Device simulation software was used to determine electrical characteristics of a given device structure for a set of bias conditions.

First is the ATHENA process simulation of the HBT and second the ATLAS electrical analysis. Process simulation for SiGe HBTs is distinguished from complementary metal-oxide-semiconductor (CMOS) process simulation by its critical dependence on Si-Si$_{1-x}$Ge$_x$/Si epitaxy and the interaction of polysilicon on silicon in the formation of the emitter and base structures. The challenge for SiGe HBT process simulation is to accurately simulate SiGe epitaxy and heavily doped polysilicon-silicon systems and combine this capability with the CMOS-specific process operations that dominate a modern SiGe bipolar complementary metal-oxide semiconductor (BiCMOS) technology.

The emergence of SiGe BiCMOS technologies has made bipolar transistor simulation an important part of the demand for state-of-the-art process simulation capability. SiGe BiCMOS technologies utilize many of the same process operations as CMOS devices and, therefore, benefit directly from the simulation expertise gained from submicron CMOS process modeling research. However, one disadvantage of adding Ge in the base is the possibility of the formation of a parasitic barrier at the base–collector junction, which becomes more pronounced at high current densities. The epitaxial base in SiGe HBTs is a complex region in which B, Ge, and carbon atoms are deposited. Point defects flow through the base as a result of P and As out-diffused from the subcollector, emitter diffusions, extrinsic base implants, and selectively implanted collector (SIC) implants, all of which influence the boron diffusion in the intrinsic base. In the SiGe or SiGeC base both the Ge and C concentrations have a strong influence in the base diffusion mechanics. It is well known that C controls the B diffusivity in SiGeC base by trapping the interstitials, while the Ge concentration affects the B diffusion by both strain and Ge:B clustering mechanisms. The major SiGe HBT fabrication process steps are:

- o Isolation trench definition
- o Subcollector implantation
- o Subcollector activation
- o Polysilicon base–contact deposition
- o Definition of base opening
- o SiGe base deposition
- o Collector implantation (SIC)
- o Nitride spacer definition
- o Metallization and final anneal

The ATHENA run starts by defining the mesh and the silicon substrate. A 2D SiGe HBT structure is constructed using a technology computer-aided design (TCAD) process simulator ATHENA. The process started with the formation of the n^+ buried collector layer by ion implantation of As ions at 50 keV at a dose of 7×10^{17} cm^{-3}. This was followed by a diffusion drive-in at 1100°C for two minutes. An arsenic-doped, 0.1 μm thick collector epitaxial layer was formed next, followed by the formation of the shallow trench isolation (STI) regions and the n^+ collector reach through by ion implantation using arsenic. Geometrical etching was used to form the STI regions, which implied that the silicon etched was merely removed from the structure. In addition, geometrical etching, which is simulated as a low-temperature process, ignores impurity redistribution. Note that geometrical etching was used to simulate all etching steps in this SiGe HBT fabrication process. The base region was fabricated next.

The extrinsic p^+ base polysilicon and appropriate etch stop layers are deposited and the emitter window is opened. Next, the SIC is fabricated using the extrinsic base polysilicon as a self-aligned mask. An under etch is performed to remove a thin layer of silicon under the emitter window. The intrinsic base had a thickness of 20 nm and a uniform boron doping profile of 1×10^{19} cm^{-3}. The next step is the deposition of the silicon-germanium layer. The Ge content was graded from 10% at the emitter side to 30% at the collector side. To reduce boron out-diffusion during subsequent thermal cycles, a boxed carbon profile of 2 × 20 cm^{-3}, which corresponds to 0.4% concentration, was also incorporated into the SiGe base. The emitter polysilicon was deposited followed by a thermal anneal. Out-diffusion of arsenic from the emitter polysilicon formed the monoemitter region. Note that in the latest fabrication processes, both the emitter and base polysilicon layers are silicided with titanium, cobaltm or nickel to reduce their sheet resistivity. The cross section of the ATHENA-constructed device is shown in Fig. 5.4a. The doping profile is shown in Fig. 5.4b. Note that the 2D SiGe HBT structure constructed is fairly ideal.

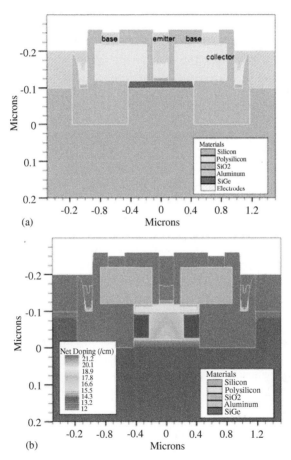

Figure 5.4 (a) Cross section of a simulated SiGe HBT. (b) Doping profile in the simulated SiGe HBT.

Quite often, the mesh used in process simulations is not suitable for device simulations. In this particular case of SiGe HBTs, for example, a fine mesh is required to properly account for the implantation and the subsequent dopant out-diffusion due to annealing of the subcollector region. Also, a fine mesh is needed to properly resolve the implantation of the SIC region. However, from the device simulation point of view, a fine mesh is generally only required for the emitter and base regions. Since in most cases the simulation time grows geometrically with the number of nodes, having a mesh that is too fine unnecessarily increases the simulation time. DevEdit, which is a software program part of the Silvaco simulation software suite, is used to remesh the structure obtained from ATHENA for the device simulations.

The new mesh for the structure ensures a gradual change in the mesh from the coarse regions to the fine regions, which is also important for numerical convergence. The final structure showing the new mesh is shown in Fig. 5.5a,b. The following models and options are enabled in the device simulations: drift-diffusion equations, lattice heating, Auger recombination, concentration-dependent Shockley–Read–Hall recombination, concentration-dependent mobility, field-dependent mobility, band-to-band tunneling, bandgap narrowing, and Fermi–Dirac statistics. Several important SiGe HBT parameters such as doping profile, Ge content, band diagram, Gummel plot, and cutoff frequency are shown in Figs. 5.6 through 5.10.

For radio-frequency (RF) applications, advanced simulations of high-frequency noise as well as intermodulation characteristics may be useful in the intrinsic device design phase but are often not considered necessary for predictive modeling. RF noise simulations are possible using impedance field methods. The noise simulation based on the impedance field method module in ATLAS was employed for the noise simulation. The correlation between the base and collector terminal noise sources is captured. Figure 5.11 shows the NF_{MIN} at 1.9 GHz versus collector current density of the simulated SiGe HBT.

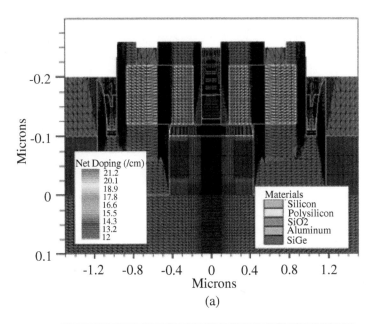

(a)

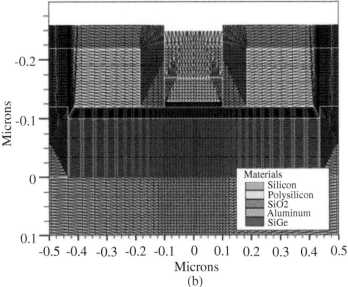

(b)

Figure 5.5 (a) Cross section of the simulated SiGe HBT structure after remeshing. (b) Cross section of the simulated SiGe HBT structure after remeshing (zoomed-in view).

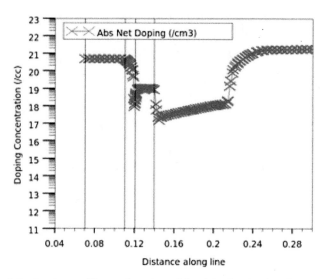

Figure 5.6 Doping profile as a function of depth in the simulated SiGe HBT.

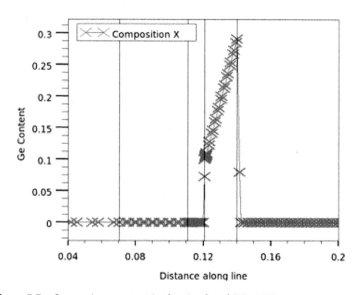

Figure 5.7 Germanium content in the simulated SiGe HBT.

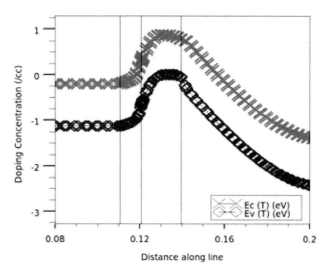

Figure 5.8 Band diagram of the simulated SiGe HBT.

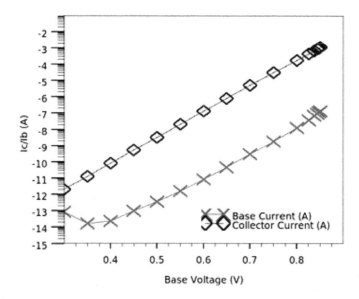

Figure 5.9 Gummel plot of the simulated SiGe HBT.

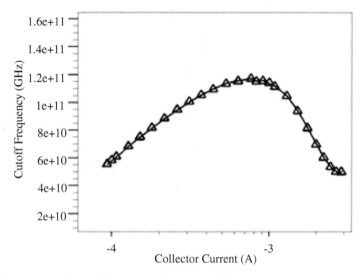

Figure 5.10 Cutoff frequency f_T versus collector current density of the simulated SiGe HBT.

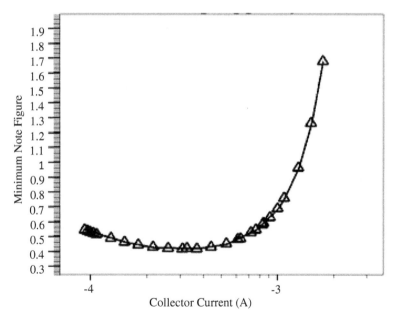

Figure 5.11 NF$_{MIN}$ at 1.9 GHz versus collector current density of the simulated SiGe HBT.

5.2   High-Speed SiGe HBTs

Great motivation for continued performance scaling exists as a result of increasing performance requirements for existing RF through millimeter-wave applications, as well as emerging applications such as millimeter-wave to submillimeter-wave radars and sensors for security, automotive, and medical applications. The wide spectrum of applications includes radar to automotive (77 GHz), wireless broadband network, WiHDMI (60 GHz), and a range beyond 100 GHz. The faster growth of the RF wireless communications market requires high-performance devices at low cost, because transistors fabricated on silicon offer ultralarge-scale integration capability and a high cutoff frequency. Figure 5.12III and Fig. 5.13IV show the roadmap predicted by the ITRS in terms of transit frequency (f_T) and maximum oscillation frequency by comparing the silicon components (CMOS and SiGe HBT) and III–V (InP high-electron-mobility transistor [HEMT], InP HBT, GaAs and GaAs pHEMT, and nHEMT).

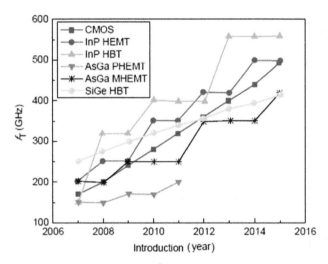

Figure 5.12   Evolution of transit frequency f_T versus introduction year for silicon and III–V technology [4].

In millimeter-wave applications, the III–V technologies have dominated the silicon technologies for many years. Performances of silicon technologies are still lower than GaAs, pHEMTs, InP, and InP HBT HEMTs. However, advancements in silicon technology driven by high-performance digital applications offer advantages to the millimeter-wave designer. With geometry scaling and technology enhancements in both CMOS and SiGe HBTs, SiGe BiCMOS technology is the most advanced on the 130 nm node and below. For example, IHP Microelectronics announced f_T/f_{max} = 300/500 GHz for its SG13G2 technology and STMicroelectronics f_T/f_{max} = 300/400 GHz for its B5T technology. IBM announced the introduction of 9HP technology SiGe BiCMOS with a couple f_T/f_{max} 300/420 GHz on the node 90 nm CMOS technology, thus further increasing the degree of integration.

In 2010, IHP Microelectronics announced a SiGeC HBT with a maximum oscillation frequency of 500 GHz under the DOTFIVE project, which is considered as the highest for SiGe devices till date. The follow-up project DOTSEVEN is targeting the development of SiGeC HBT technologies with an f_{max} of around 700 GHz, which will open the door for terahertz (THz) applications. In recent years, the development of the SiGe HBT has progressed rapidly. In terms of f_T and f_{max}, SiGe HBTs have advanced from 100 GHz of a few years ago to today's 500 GHz and beyond in several research laboratories, as shown in Fig. 5.14.

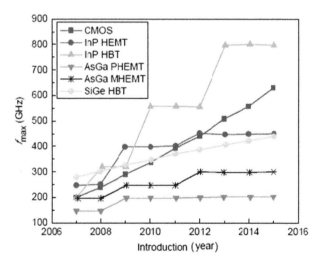

Figure 5.13 Evolution of maximum oscillation frequency f_{max} versus introduction year for silicon and III–V technology [4].

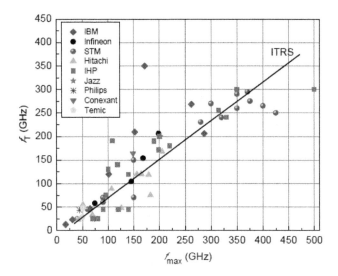

Figure 5.14 f_{max} and the associated f_T data from selected SiGe HBT vendors [4].

With the SiGe HBT breakdown voltage becoming a key challenge to performance scaling as a result of the inherent trade-off between peak f_T and breakdown voltage, the collector doping profile and Ge retrograde in the CB junction must be carefully designed. Moreover, the development of next-generation SiGe HBTs will require implementation of new structures to minimize base resistance and CB capacitance.

Using 2D device simulations, it is predicted that the cutoff frequencies of SiGe HBTs can be scaled beyond 500 GHz. These devices have the potential to enable advanced millimeter-wave circuits. However, shot noise correlation, which is captured through noise transit time, becomes increasingly important as circuit designers continue to push the operating frequencies of the circuits. Next, we investigate the technology development of state-of-the-art SiGe and SiGeC HBTs by means of TCAD. The objective of this work is to obtain an advanced HBT device very close to the real device, not only in its process fabrication steps, but also in its physical behavior, geometric architecture, and electrical results. This investigation may lead to achieve the best electrical performances for the devices studied, in particular a maximum operating frequency of 500 GHz.

In the following sections, important parameters of SiGe HBTs (f_T, f_{max}, and V_A) will be considered in detail and attempts are made to illustrate how simulation has been used to optimize the device design for circuit applications. Base, emitter, and collector profile design issues at room temperature will be discussed. All the simulations have been performed using the Sentaurus device simulator. The description of the device technology named B3T is done, illustrated with structures obtained from process simulation. Models used for diffusion of doping species versus layer composition are discussed, followed by the introduction of the simulation process flow. The methodology used to calibrate model parameters is exposed and simulation results are compared to measurements made in real devices. The impact of some technological parameters (boron doping level, carbon and germanium contents) on doping profiles or on electrical performances is analyzed. Finally, simulation results are compared to fabricated B3T devices.

The predictive simulation of a SiGeC HBT with a base biaxially stressed on silicon requires taking into account the combined effects of specific properties of the ternary SiGeC alloy and the biaxial strain induced by the lattice mismatch. In our simulations, we consider anisotropic analytical mobility models based on Monte Carlo simulations for a low electric field for SiGeC alloys.

Of all the parameters studied, in this study we will focus on the influence of the boron doping level in the base, the carbon content, and germanium on the device characteristics. These three parameters are very important for the improvement of transistor performance. On the percentage carbon for obtaining good performance for the transistor, it can be concluded that it must remain below 1%, with an optimum of about 0.5% for maximum-possible f_T and f_{max}.

5.3 SiGeC HBTs: Process and Device Simulation

We consider the SiGeC bipolar process technology in the latest high-speed 0.13 μm BiCMOS technology from STMicroelectronics, as available in Ref. [3]. The development of this TCAD simulation is based on STMicroelectronics' advanced BiCMOS technology for SiGe and SiGeC HBTs. The front-end-of-line process of SiGeC HBT fabrication is shown in Fig. 5.15. The main steps of this HBT fabrication process is also illustrated in Figs. 5.16 through 5.24, with device cross sections obtained from the process simulation. The architecture and process flow of the HBT is quite complex due to diffusion of doping species (boron, arsenic) in the strained-SiGeC layer and formation of polycrystalline materials (emitter, extrinsic base layer) during the process. On a p-type silicon substrate, the process starts with the deposition of a thin oxide layer to keep out dirt, followed by an arsenic implantation to form the n+ buried layer. After the removal of the oxide a lightly doped n-type layer is deposited to form the collector. The active area of the transistor is then isolated and defined by the formation of the STI (Fig. 5.17). To contact the collector buried layer and to avoid a high-resistivity area between the buried layer and the contact area, n^+ implantation (SINKER) is performed (Fig. 5.17). This is followed by the CMOS well implantation, the definition and opening of the bipolar area. The pedestal oxide layer is deposited, followed by a P-doped polybase and a stack of oxide/nitride layers (Fig. 5.18). Once etched selectively, the pedestal oxide is used to prepare the intrinsic base region and the connection with the extrinsic polybase layer.

After these successive depositions the emitter window is defined and opened via photolithography (Fig. 5.19). Thin and thick layer stacks of oxide and nitride are used and anisotropically etched. The pedestal oxide is isotropically etched in order to release the surface on top of the collector, giving enough space under the polybase for the subsequent connection between the intrinsic and the extrinsic base (Fig. 5.20). The fully self-alignment of junctions lies on this method associated with the selective SiGeC base epitaxy. The SIC is carried out through this opening. The boron-doped SiGeC base is then grown epitaxially, producing a thin single-crystal layer on top of the SIC (Fig. 5.21). The base growth is followed by the emitter spacers' formation (Fig. 5.22) and the polyemitter deposition (Fig. 5.23). The final device is defined by the polyemitter and polybase patterning (Fig. 5.24) and finalized with the spike activation and the cobalt silicidation. Then, tungsten plugs and via are processed, and the subsequent BiCMOS 6 levels metal back-end-of-line is performed.

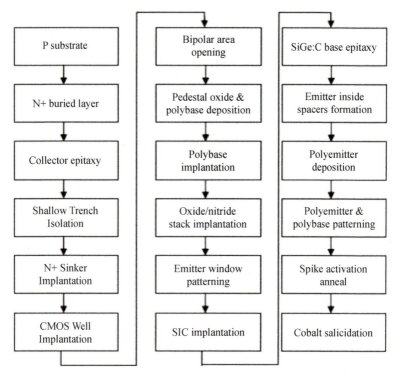

Figure 5.15 Process flow for SiGeC HBT fabrication. Main fabrication steps are shown [3].

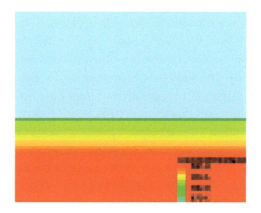

Figure 5.16 Implantation of the N + buried layer and collector epitaxy. This figure and the following figure show the cross section of the HBT process.

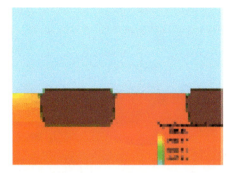

Figure 5.17 Shallow trench isolation (STI in dark region) and N+ sinker implantation (between the two STI). The left edge of the figure is the central axis of the HBT.

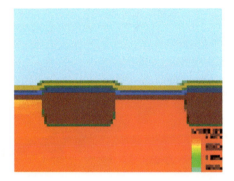

Figure 5.18 Deposition of the pedestal oxide (dark), the polybase, and the nitride.

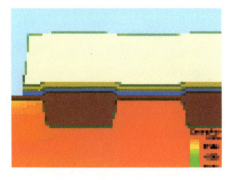

Figure 5.19 Emitter window patterning (thick protecting layers, anisotropic etching).

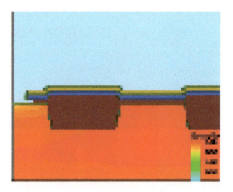

Figure 5.20 Oxide etching to define the intrinsic base region (self-alignment of the two junctions) and selective implantation of the collector (SIC) owing to this nitride uncovered region (left side of the figure).

Figure 5.21 Intrinsic SiGe:C boron-doped base epitaxy, and Si-cap.

Figure 5.22 Emitter inside spacers' formation.

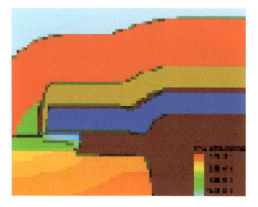

Figure 5.23 Polyemitter depositions and annealing.

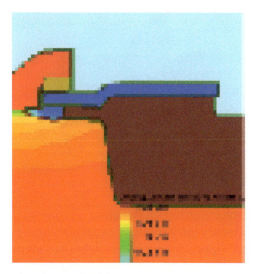

Figure 5.24 Polyemitter and polybase patterning.

Figure 5.25 illustrates the 2D device structure generated with the TCAD tool. As detailed in previous sections, the HBT results from a double-polysilicon fully self-aligned (FSA), with a boron-doped SiGeC base and one arsenic-doped monoemitter. The collector module is formed by an n+ buried layer, a collector epitaxy, and a SIC. Figure 5.26 focuses on the core of the HBT, showing more finely the different region of Si, SiGeC, oxide, nitride, and poly.

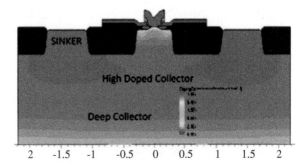

Figure 5.25 Structure generated with the 2D simulation. Brown areas indicate oxide regions, orange areas the n-type doping, and blue areas the p-type doping. Because of symmetry, only one half of the structure is simulated [3].

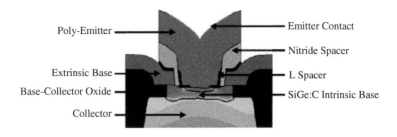

Figure 5.26 Structure obtained at the end of the electrical simulation with its main components. Brown areas indicate oxide regions, orange areas the n-type doping, and blue areas the p-type doping [3].

In the Sentaurus process simulator (SProcess), the different fabrication steps are introduced in an *Sprocess command file.* This file is the main input for the process simulator, containing all the process steps, and it can be edited according to user specifications. One-dimensional simulations are performed in order to obtain accurate layer profiles of dopants under the emitter window. The base epitaxy is formed by a Si/SiGeC/Si stack with a total thickness of about 50 nm. Boron doping and carbon incorporation is done during the epitaxy only in the central part of the SiGe layer. Ge content follows a graded profile, with 20% to 30% from the emitter side to the collector side of the SiGe base. The SiGe thickness is typically 20 nm. During the growth and the final thermal annealing, the diffusion of doping species gives broader layers than those deposited. Figure 5.27 shows the final structure obtained from TCAD according to the technology process flow shown in Fig. 5.15. The 2D process simulations lead to an accurate representation of real device and consequently realistic results. Figure 5.27 shows such critical parameters as L-shaped spacer length and thickness, nitride spacer thickness, BC oxide thickness, and external-to-internal base connection length. The 1D and 2D simulations are gathered for the species and doping profiles, and this structure will be used for device simulation.

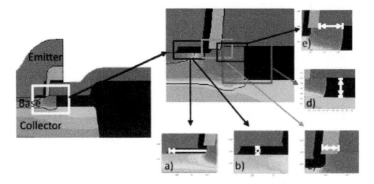

Figure 5.27 2D key process parameters: (a) L-shaped oxide spacer width, (b) L-shaped oxide spacer thickness, (c) nitride spacer, (d) base–collector oxide, and (e) external-to-internal base connection [3].

Boron diffusion effect on the base width is shown in Fig. 5.28. The 2D representations shown in Fig. 5.28 illustrate the behavior for a doping level of 2×10^{19} cm^{-3} ($\sim 7 \times 10^{18}$ cm^{-3} after diffusion), for which the base width is about 15 nm (Fig. 5.28b); for a doping level of 5×10^{19} cm^{-3} ($\sim 1 \times 10^{19}$ cm^{-3}), this is about 20 nm (Fig. 5.28c), and for a doping level of 8×10^{19} cm^{-3} ($\sim 2 \times 10^{19}$ cm^{-3}), this is about 25 nm (Fig. 5.28d). Effects of carbon diffusion on the base width are shown in Fig. 5.29. The 1D doping profile of Fig. 5.29a shows the reduction of the boron diffusion with the addition of carbon. For the same B doping level (5×10^{19} at/cm^{-3}), the increase of the C content leads to a higher boron peak and to a narrower base. Figure 5.29 shows this effect for three different contents of carbon. Thus a content of 0.01% leads to a base width of 28 nm (Fig. 5.29b), while 0.05% leads to 25 nm (Fig. 5.29c), and 0.1% reduces the base width to 20 nm (Fig. 5.29d).

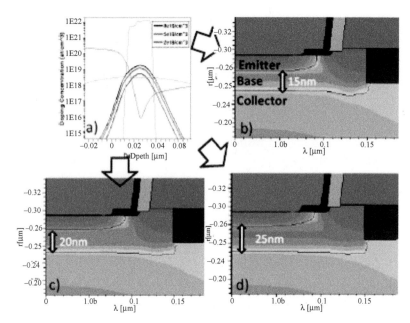

Figure 5.28 (a) Simulated 1D profile for three different boron doping levels, (b) base width resulting from a doping level of 2×10^{19} at/cm^3, (c) base width for B doping of 5×10^{19} at/cm^3, and (d) base width for B 8×10^{19} at/cm^3 [3].

After the 1D and 2D process simulation toward a predictive TCAD simulation, the device simulations are performed to account for the impact of the strained-SiGeC layer on the different electrical characteristics of the device. Figure 5.30 shows the structure simulated electrically, with a meshing density adapted to the HBT. The main material models used in all the device simulations are the Shockley–Read–Hall doping-dependent lifetime recombination model, Caughey–Thomas doping-dependent mobility model, and electric field–dependent mobility model.

For accurate device calibration, many electrical characteristics are compared to measurements, and a calibration methodology is applied with the following electrical characteristics:

- The Gummel plot in order to precisely reproduce the collector and base currents in the low-to-moderate injection regime
- The standard Gummel plot for the high injection regime

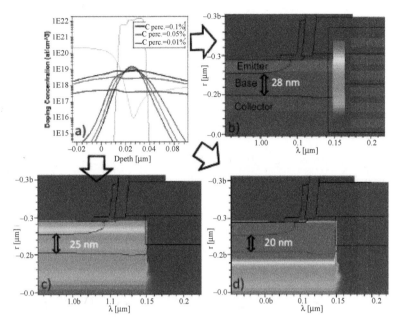

Figure 5.29 Impact of the carbon content on boron diffusion and base width. (a) 1D representation for three different boron doping levels and (b) 2D simulation for C percentage = 0.1%, (c) for C percentage = 0.05%, and (d) for C percentage = 0.01% [3].

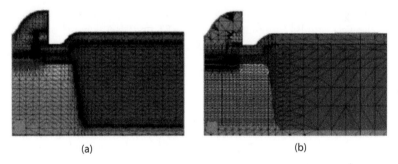

Figure 5.30 TCAD cross section of a SiGe HBT elaborated under B3T technology process flow. (a) Process meshing and (b) electrical meshing of the final structure. 1D and 2D simulations are gathered for device simulation [3].

- The $I_b(V_{CE})$ characteristics extracted with the BV_{CEO} simulation for the collector resistance and the base–collector junction

- The $f_T(I_c)$ characteristics for the good evaluation of the dynamic device parameters
- The $f_{max}(I_c)$ characteristics for the good evaluation of the dynamic device parameters and parasitic capacitances and resistances

Next, we discuss different possible architectures of advanced SiGeC HBTs for achieving an f_{max} of 500 GHz. The simulated devices were fabricated in the framework of the Dotfive European program, where the objective was to design HBTs with a maximum oscillation frequency of 500 GHz. The simulated devices are SiGeC HBTs fabricated in low-cost B3T architecture and B4T and B5T technologies. B3T is a bipolar process technology derived from high-speed 0.13 µm BiCMOS technology from STMicroelectronics. It features 260 GHz/350 GHz for f_T/f_{max}. The base epitaxy is formed by a Si/SiGeC stack with a thickness of about 40 nm of which 20 nm is the SiGeC base. The Ge-graded profile increases from the emitter side up to the collector side, varying from 20% to 30%. The C content is less than 1% within the same region. B4T was proposed to develop a device to give maximum oscillation frequencies beyond 400 GHz. This is an FSA HBT with a selective epitaxial growth of the boron-doped SiGeC base layer. The collector is formed by an n$^+$ buried layer, an epitaxial low-doped layer on it, and an SIC. The emitter is arsenic doped and covered by the polyemitter. The B5T technology is based in the B3T and B4T device platforms. B5T platform was proposed to improve the performance by pushing to limits of processing steps and find an optimized technology with design of experiments. The increase in frequency of operation can be obtained by optimizing the geometry doping profiles in the transistor regions and by reducing parasitic elements.

The main key process parameters and frequency values of the devices studied are compared in Table 5.1. The equivalent models used by default in the device simulator for the SiGeC HBT device are as follows:

- Klaassen's mobility model, which unifies the description of majority and minority carrier silicon bulk mobilities. The model includes the temperature dependence of the mobility and takes into account electron–hole scattering, screening of ionized impurities by charge carriers, and clustering of impurities.

- The carrier energy relaxation time is assumed by the simulator as the same value as that of bulk silicon. By default, this is constant and is set to 0.3 ps for electrons and 0.25 ps for holes.
- The bandgap model with Ge mole fraction dependencies. The f_T and f_{max} performances for B5T devices are compared in Fig. 5.31 with B4T and B3T technologies. While the f_T enhancement is moderate among the three transistor technologies with a maximum value of ~260 GHz for B3T and B4T+ and ~280 GHz for B5T, f_{max} is improved by 17% from B3T to B4T+ and by 18% from B4T+ to B5T. Figure 5.32 compares f_{max} and f_T values for the three architectures studied. The three structures have the same emitter window (200 nm) and the same emitter junction width (110 nm). A comparison of dynamic performances for the three architectures is made.

Table 5.1 Comparison of process parameters and frequency values for the technologies studied

	B3T	B4T+	B5T	Units
N+ buried layer doping	4×10^{17}	8×10^{17}	8×10^{17}	at/cm^{-2}
Emitter window	250	230	170	nm
Emitter width (W_E)	120	90	75	nm
SIC dose/energy	$6 \times 10^{13}/230$	$6 \times 10^{13}/155$	$5 \times 10^{13}/155$	cm^{-2}/keV
Si cap	18	15	14	nm
Carbon	0.8	0.8	0.5	%
Boron doping	5×10^{19}	5×10^{19}	5×10^{19}	at/cm^{-3}
f_T	260	275	295	GHz
f_{max}	350	425	554	GHz

The 1D key process parameters are:

- Silicon cap thickness implantation
- Ge percentage, collector side
- Ge percentage in the middle
- Ge percentage, emitter side
- C percentage
- Emitter doping

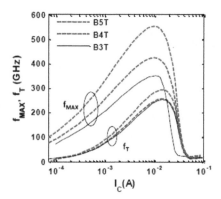

Figure 5.31 TCAD simulations: comparison of f_T and f_{max} versus I_c for B3T, B4T, and B5T technologies for $V_{CB} = 0.5V$ at 300 K ($V_{CE} \approx 1.35V$) [3].

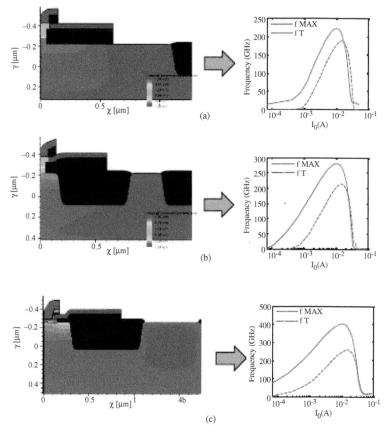

Figure 5.32 Comparison of dynamic performances for the three architectures. (a) Low-cost B3T, (b) buried collector B3T, and (c) B4T at $V_{CB} = 0$ V [3].

The 2D key process parameters are:

- Emitter window opening
- Base–collector oxide thickness
- Emitter width
- SIC option
- SIC implantation energy
- Definition of SIC depth

5.4 Strain-Engineered SiGe HBTs

The introduction of silicon-germanium in the base region of silicon-based bipolar transistors offers improved overall performance at higher operating frequencies. SiGe BiCMOS technology features important advantages such as:

- Inherently high forward current gain
- Homogeneous high integration of bipolar transistors and CMOS
- Power saving due to a higher transit frequency allowing a lower bias current for a given gain at a given frequency
- Lower noise and high linearity

Advances in the development of Si/SiGe heterostructure devices have revealed many exciting possibilities for extending CMOS and BiCMOS technologies comfortably into RF circuitry. The current state-of-the-art SiGe BiCMOS is based on the integration of an advanced CMOS with n-p-n SiGe base HBTs on a silicon substrate. Consequently, the development and optimization of HBT devices has been mainly focused on n-p-n HBTs.

Currently the use of SiGe HBTs is widespread in CMOS technology. For over the last 50 years, performance enhancement in Si CMOS technology has been via the scaling laws. However, below the 90 nm technology node, performance enhancement through device scaling, such as shrinking the gate length and thinning the gate oxide, has become more and more difficult because of several physical limitations in downscaling of metal-oxide-semiconductor field-effect transistors (MOSFETs). The so-called technology boosters as per the International Technology Roadmap for Semiconductors (ITRS) 2013 edition include strained-Si channels, ultrathin silicon-on-insulator (SOI), metal gate electrodes, multigate structures, ballistic transport channels, metal source/drain junctions, etc. Among the technology boosters, strained-Si channels (see Ref. [1] of Chapter 1) have been recognized as the technology applicable to near-term technology nodes. Strain engineering involves uniaxial and biaxial strain induction in silicon wafers. Process-induced strain and substrate-induced strain have become the main avenues for strain engineering. The key differences between process-induced and substrate-induced strain have been discussed (see Ref. [10] of Chapter 1). Rapid progress in local strain techniques has been included in 90 nm logic CMOS technologies and below. Advancements in silicon technology driven by high-performance digital applications offer advantages to the millimeter-wave designer. Performance, quantified by f_T and f_{max} has dramatically increased with geometry scaling and technology enhancements in both CMOS and SiGe HBTs. High-performance silicon-germanium technology has been used in high-frequency transceivers for millimeter-wave applications. To maintain this rapid rate of improvement, aggressive strain engineering of bipolar devices is required.

SiGe HBTs are realized by the pseudomorphic growth of the SiGe layer on a Si substrate, which allows engineering of the base region to improve performance where the base has a smaller energy bandgap than the emitter, which increases the gain. The energy bandgap of SiGe decreases with increasing Ge mole fraction, but the maximum Ge content is limited by the amount of strain that can be accommodated within a given base layer thickness. Therefore, a new innovation is necessary to overcome this limitation and meet the continuous demand for high-speed devices. Reports on attempts toward improvement of the performance of bipolar transistors using strain technology are limited. Application of the strain technology at a selected region of a bipolar device by means of a strained layer can strongly enhance the device performance due to the sensitivity of silicon material to strain.

Next, new device concepts and novel device architectures for bipolar devices that are based on strain engineering technology will be explored using TCAD modeling. The impact of strain engineering in n-p-n SiGe HBT devices on the electrical properties and frequency response will be presented. New n-p-n SiGe HBT device architecture employing strain engineering technology at the base and collector regions will be presented. In this approach the desired strain is introduced during the device fabrication process through the specific device architecture employing silicon nitride (Si_3N_4) and silicon oxide (SiO_2) strain layers at the collector region of the bipolar device. The TCAD simulation methodology used consists of the following steps:

1. Virtually fabricate the device using process simulation.
2. Incorporate strain in process and evaluate the strain level obtained in the device.
3. Study the strain distribution in different zones in the device.
4. Define simulation parameters and physical models for device simulation.
5. Perform device simulation to analyze the device electrical performance.

5.5 N-P-N SiGe HBTs with an Extrinsic Stress Layer

Process and device simulations have been performed using Sentaurus TCAD software tools to build the device structure and to calculate the associated mechanical stress. The final devices structure is exported to Sentaurus Device for device simulation. The major processing steps of the stressed n-p-n SiGe HBT device architecture are:

- Shallow and deep trenches' isolation
- Deposition of etch stop material and a thin layer of polysilicon
- Formation of a base region
- Deposition of oxide and nitride layers
- Formation of an emitter mandrel and recesses
- Deposition of a SiGe extrinsic stress layer
- Emitter opening and formation of sidewall spacers
- Formation of a T-shaped emitter

The complete HBT device structure with a SiGe extrinsic stress layer is shown in Fig. 5.33. Specific models for the SiGe bandgap, bandgap narrowing, effective mass, energy relaxation, mobility for hydrodynamic (HD), and drift-diffusion simulations have been implemented in the simulation. The study examined not only the transistor DC performance but also the RF performance through multiple optimizations for the explored vertical transistors.

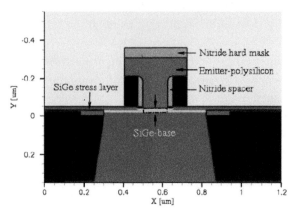

Figure 5.33 Complete structure of the n-p-n SiGe HBT device with an extrinsic SiGe stress layer [5].

Due to the addition of the extrinsic stress layer, stress is generated inside the device (i.e., S_{xx}) and the stress values are calculated at the middle of the base. Approximately 500 MPa of an additive compressive stress (S_{xx}) is generated at the base region, and 500 MPa of tensile stress (S_{xx}) is lessened at the collector region. Chapter 2 provides a detailed discussion of the stress calculation and the impact strain on the bandgap energy and the carrier mobility. Sentaurus TCAD software tools have been used to perform the 2D device simulations using an HD transport model. The default silicon models and parameter files included in the TCAD software library cannot be used for the SiGe device with external stress. Therefore, specific parameter files and physical models have been calculated and were implemented during device simulation.

To obtain accurate simulation results, the HBT device has been divided into two regions: The first one is the SiGe base, which is divided into two zones corresponding to different Ge content and doping concentration. The second region is the remaining part of the device without Ge content (the emitter and collector regions). The different device regions and zones are shown in Fig. 5.34.

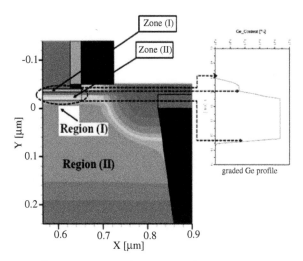

Figure 5.34 Different regions and zones used for device simulations [5].

Simulation results show that introducing an extrinsic stress layer on the HBT device structure will enhance the maximum oscillation frequency and the cutoff frequency. Approximately 5% of improvement in f_{max}, and 3% of improvement in f_T have been achieved in comparison to the standard, conventional SiGe HBT device (without an extrinsic stress layer). The intensive study of the transit time in the strain-engineered n-p-n SiGe HBT device shows a significant decrease in total transit time in comparison to the standard, conventional n-p-n SiGe HBT device, which arises from the reduction of the collector transit time, verifying that silicon material is more sensitive to strain than the SiGe base region and the device performance improvements are mainly due to the impact of the induced strain at the collector region. The components of the transit time have been extracted for both SiGe HBT devices as a function of the collector current and are shown in Fig. 5.35. Simulation results show that introducing the extrinsic stress layer on the device structure will decrease the total transit time and hence improve the device performance.

In addition, the impact of changing the device's emitter width on the device performance has been studied. The result shows that increasing the emitter width will decrease the stress values induced at the base, resulting in degradation in device performance. The impact of changing the Ge content at the extrinsic stress layer on the stress values generated inside the device (S_{xx}, and S_{zz}) has been studied. As expected, increasing the Ge content at the extrinsic stress layer will increase the stress values generated inside the device. This is related to the increase of the lattice constant difference between the SiGe base layer and the SiGe stress layer, which will increase the stress values induced at the base. Nevertheless, increasing the Ge content at the stress layer will also increase the misfit dislocations between the two layers, which may cause degradation of device performance. Therefore, the Ge content at the stress layer must be controlled and chosen carefully to ensure device performance enhancement.

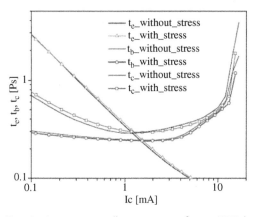

Figure 5.35 Transit time versus collector current for an HBT device with and without an extrinsic stress layer [5].

5.6 N-P-N SiGe HBT Device Employing a Si_3N_4 Strain Layer

Application of strain on the base region of an n-p-n SiGe HBT device has shown that it is less efficient in comparison to the Si bipolar junction transistor (BJT) as the SiGe base is already stressed due to the existence of Ge at the base. Here, a new n-p-n SiGe HBT device architecture employing strain engineering technology at the collector region will be presented. In this approach the desired strain is introduced during the device fabrication process through a specific device architecture employing silicon nitride (Si$_3$N$_4$) using the local strain technique. Nitride films can induce stresses greater than 1 GPa upon thermal treatment, which arises from two sources, the coefficient of thermal expansion mismatch between silicon and the nitride film and intrinsic film stress caused by film shrinkage. Processing conditions such as temperature, pressure, deposition power, and reactant and impurity concentrations are important factors in determining the magnitude and strain type (i.e., compressive or tensile). Details on the stress model used for the nitride film are available in Chapter 2. Process simulations are performed using Sentaurus TCAD software tools to build the device structure and to calculate the associated mechanical stress. The major processing steps of the silicon-nitride-stressed n-p-n SiGe HBT device architecture are:

- Deposition of a silicon layer
- Etching of the silicon layer
- Deposition of a nitride layer
- Nitride layer etching and silicon layer deposition
- Deposition of a SiGe base
- Deposition of an oxide layer, emitter opening, and deposition of a polysilicon layer
- Etching of oxide and nitride layers and formation of a T-shaped emitter

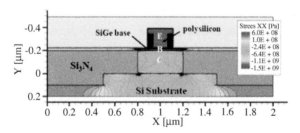

Figure 5.36 Final device structure showing stress isocontour lines generated inside the device [5].

Figure 5.36 shows the final device structure showing stress isocontour lines generated inside the device. Figure 5.37 shows a comparison between the stress values generated inside the standard, conventional SiGe HBT device and the proposed SiGe HBT device employing a nitride strain layer at the collector region. Sentaurus TCAD software tools have been used to perform the 2D device simulations using an HD model without taking in consideration self-heating effects. The stress-induced mobility enhancement has been calculated using the piezoresistivity model. A detailed description of the models used in the simulation is presented in Chapter 2.

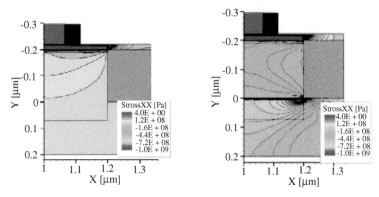

Figure 5.37 Isocontour lines representing the stress values generated inside the device; standard, conventional n-p-n SiGe HBT device (left) and strained–silicon nitride n-p-n SiGe HBT device (right) [5].

Simulation results of a comparison of f_T and f_{max} for a strained-HBT and a standard, conventional SiGe HBT device (without strain) are shown in Fig. 5.38 as a function of the collector current. The values of f_T and f_{max} are obtained assuming a constant-gain-bandwidth product (–20 dB/decade slope) with respect to the current gain $|h_{21}|$ and the unilateral gain $|U|$ characteristics at a spot frequency of 30 GHz. The influence of introducing a nitride strain layer at the collector region is demonstrated. The results show that the poststrain HBT device exhibits better high-frequency characteristics in comparison to an equivalent standard, conventional device. Approximately 8% of improvement in f_T and 5% of improvement in f_{max} have been achieved for the strained-silicon n-p-n SiGe HBT device. A trade-off between f_T and f_{max} values may be achieved, as shown in Fig. 5.39, where f_T is plotted versus f_{max} for different BV_{CEO} values (i.e., with different collector doping). As shown in Fig. 5.39, an improvement of up to 43% in the f_T value can be achieved for a given f_{max}, and up to

7% of improvement in the f_{max} value for a given f_T can be achieved by means of strain engineering, as well by choosing the proper collector doping level.

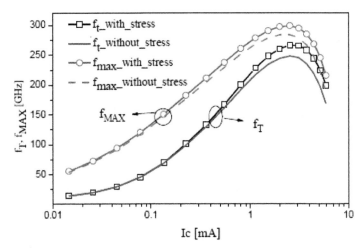

Figure 5.38 f_T and f_{max} versus I_c for both strained- and unstrained-SiGe HBT devices (W_E = 130 nm, V_{BC} = 0 V) [5].

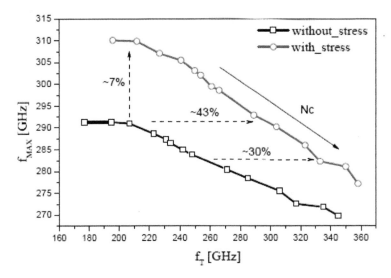

Figure 5.39 Trade-off between f_T and f_{max} for both strained- and unstrained-SiGe HBT devices [5].

5.7   N-P-N SiGe HBT Employing a SiO$_2$ Strain Layer

A second approach to improve the SiGe HBT device performance is to introduce strain in the device through using a silicon oxide (SiO$_2$) stain layer at the collector region during device fabrication. Silicon has a high affinity for oxygen, and an amorphous native oxide film rapidly forms on Si upon exposure to an oxidizing ambient. During thermal processing the SiO$_2$ layer expands and contracts at different rates compared to the silicon substrate according to their thermal expansion coefficients. Because of this mismatch, as well the growth of oxide on top of the silicon substrate, a mechanical strain is induced. In this case, the desired strain is induced at the collector region. A mechanical compressive strain is induced along the horizontal axis, and a mechanical tensile strain is induced along the vertical axis. Like the n-p-n SiGe HBT device employing a nitride strain at the collector region, process simulations are performed using Sentaurus TCAD software tools to build the device structure and to calculate the associated mechanical stress. The major processing steps are:

- Deposition of a silicon layer
- Etching of the silicon layer
- Deposition of oxide, etching of oxide, and deposition of polysilicon
- Deposition of silicon, followed by etching of oxide and polysilicon layers
- Deposition of a SiGe base
- Deposition of an oxide layer, emitter opening, and deposition of a polysilicon layer
- Etching of oxide and nitride layers and formation of a T-shaped emitter

Figure 5.40 shows the final device structure obtained from process simulation showing stress isocontour lines generated inside the device. Figure 5.41 shows a comparison of isocontour lines representing the stress values generated inside the device: a conventional n-p-n SiGe HBT device (left) and a strained-SiO$_2$ n-p-n SiGe HBT device (right). TCAD Sentaurus software tools have been used to perform the 2D device simulations using an HD model without taking in consideration the self-heating effect and thermal behavior. However, the device's thermal behavior could be degraded by the formation of the oxide layer through affecting the device's thermal conductivity. The f_T and f_{max} characteristics as a function of the collector current for an n-p-n SiGe HBT device utilizing a SiO₂ strain layer at the collector region and a standard, conventional n-p-n SiGe HBT device are shown in Figs. 5.42a and 5.42b, respectively.

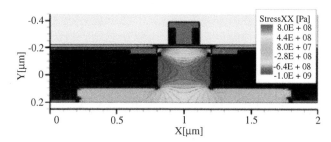

Figure 5.40 Final device structure obtained from process simulation showing stress isocontour lines generated inside the device [5].

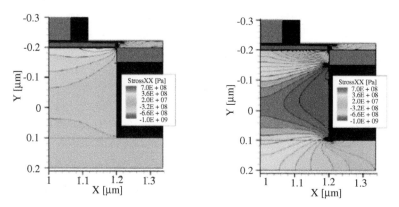

Figure 5.41 Isocontour lines representing the stress values generated inside the device; standard, conventional n-p-n SiGe HBT device (left) and SiO₂ strained n-p-n SiGe HBT device (right) [5].

The impact of introducing a SiO₂ strain layer at the collector region on the device frequency response is demonstrated. Approximately 14% of improvement in f_T, and 9% of improvement in f_{max} have been achieved for n-p-n SiGe HBT device architecture employing a SiO_2 strain layer at the collector region in comparison to an equivalent standard, conventional HBT device. A trade-off between f_T and f_{max} values may be achieved, as shown in Fig. 5.43, where f_T is plotted versus f_{max} for different BV_{CEO} values (i.e., with different collector doping). As shown in Fig. 5.43, an improvement up to 47% in the f_T value can be achieved for a given f_{max} and up to 14% of improvement in the f_{max} value for a given f_T can be achieved by means of strain engineering, as well by choosing the proper collector doping level.

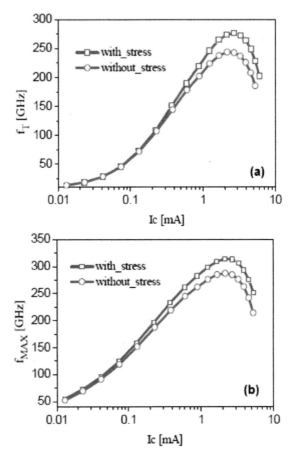

Figure 5.42 (a) f_T versus I_c and (b) f_{max} versus I_c for both strained- and unstrained-SiGe HBT devices ($V_{CB} = 0$) [5].

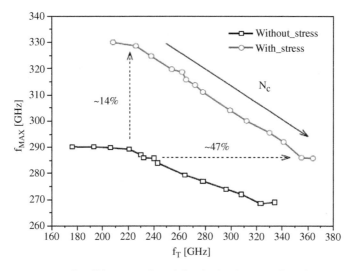

Figure 5.43 Trade-off between f_T and f_{max} for both strained- and unstrained-SiGe HBT devices [5].

5.8 Summary

In this chapter we have demonstrated the impact that TCAD can have in SiGe technology development, with an emphasis on the ability to simulate the full development cycle from the most sensitive process variation to the development of predictive compact models. The operating principle and properties of SiGe heterojunction and Si homojunction bipolar technologies are discussed from both DC and AC perspectives. The work presented in this chapter should help to obtain more physical and realistic simulations, better understanding the development and optimization of SiGe and SiGeC HBT devices. Important figures of merit such as cutoff frequency, f_T; maximum oscillation frequency, f_{max}; and breakdown voltages are discussed. As part of the simulation study, different design issues of the emitter, base, and collector are discussed. New device concepts and novel device architectures for bipolar devices that are based on strain engineering technology have been explored using TCAD modeling. The impact of strain engineering technology in n-p-n SiGe HBT devices on the electrical properties and frequency response has been studied in detail. Simulation results show that the strained-SiGe HBT device exhibits better high-frequency characteristics in comparison to an equivalent standard, conventional SiGe HBT device.

References

1. Krishna, V. R. (2009). *Physical Modelling of Advanced SiGe Transistors*, PhD thesis, University of Bremen.

2. Moen, K. A. (2012). *Predictive Modeling of Device and Circuit Reliability in Highly Scaled CMOS and Silicon Germanium BiCMOS Technology*, PhD thesis, Georgia Institute of Technology.

3. Quiroga, A. (2013). *Investigation and Development of Advanced Si/SiGe and Si/SiGeC Heterojunction Bipolar Transistors by Means of Technology Modeling*, PhD thesis, Universite Paris-Sud.

4. Sahoo, A. K. (2012). *Electro-Thermal Characterizations, Compact Modeling and TCAD Based Device Simulations of Advanced SiGe:C BiCMOS HBTs and of Nanometric CMOS FET*, PhD thesis, Universite Bordeaux I.

5. Al-Sadi, M. (2011). *TCAD Based SiGe HBT Advanced Architecture Exploration*, PhD thesis, Universite Bordeaux I.

Chapter 6

Silicon Hetero-FETs

Silicon technology has reached a very mature stage at which incremental progress is achieved with increasing difficulty. Further improvements in the performance of silicon ICs can be expected by using novel materials and device structures. The conventional way of enhancing the performance of field-effect transistors is to scale down the gate length. Another technique is to use a material with better transport properties as the conductive channel between source and drain. Solutions to achieving a higher carrier mobility and saturation velocity have been sought among different materials intended for the MOSFET channel. Process-induced strain as discussed in previous chapters is the widely adopted method for CMOS transistor. However, process-induced strain has a drawback due to its dependence on geometry and gate pitch. The performance deteriorates for gate feature beyond 45 nm. $Si_{1-x}Ge_x$ with a biaxial compressive strain has been demonstrated to be favorable for hole confinement and hole mobility enhancement, because of its band offset and split in the valence band. SiGe is the ideal material to boost the speed of both n- and p-channel Si MOSFETs.

One of these possibilities is being offered by an alloy of silicon and germanium, SiGe, grown on Si substrates, which can form the basis for a host of new high-speed electronic and optoelectronic devices (see Ref. [8] of Chapter 1). For the process-induced strain technology, the fundamental material properties and increasing process

Introducing Technology Computer-Aided Design (TCAD): Fundamentals, Simulations, and Applications
C. K. Maiti
Copyright © 2017 Pan Stanford Publishing Pte. Ltd.
ISBN 978-981-4745-51-2 (Hardcover), 978-1-315-36450-6 (eBook)
www.panstanford.com

integration complexities are some of the limitations and challenges seen by this technology. While compatible with existing silicon technology, SiGe offers a significant increase in device performance due to its superior transport properties as compared to silicon. The energy bandgap of SiGe is smaller than that of silicon and linearly scales with the Ge content, opening up exciting opportunities to build devices based on bandgap-engineering concepts. High-speed Si/SiGe CMOS devices would also facilitate a significant increase in the digital data rate. Dual-channel architecture, consisting of strained $Si/Si_{1-x}Ge_x$ on a relaxed-SiGe buffer provides a platform for fabricating MOS transistors with high drive currents, resulting from high carrier mobility and carrier velocity, due to the presence of compressively strained-silicon-germanium layer. The MOSFET structures described in this chapter use SiGe film thicknesses within the Mathews–Blakeslee critical thickness.

Most of the experimental work on heterostructure Si-SiGe MOSFETs has been carried out on devices with channel lengths greater than 0.25 μm. It is critical for the future prospects of Si-based heterostructure ICs that their performance edge be maintained as the dimensions shrink to 0.1 μm and beyond. In the deep sub micrometer realm, nonlocal transport effects play a significant role, and a meaningful attempt to predict device behavior ultimately requires their inclusion. To date, such studies have been hindered by the complexity of the numerical modeling problem and by the scarcity of experimental data referring to the transport properties of SiGe. The purpose of this chapter is to (a) propose a fully compositionally graded SiGe channel concept as leading to the best performance in deep sub micrometer MOS devices, (b) quantify the high-frequency performance of Si-SiGe FET's, and (c) estimate the impact of compositional channel grading on hot-carrier and short-channel effects.

6.1 Electronic Properties of Strained Si and SiGe

Silicon and germanium are column IV semiconductors with a diamond crystal structure. The mismatch between their lattice constants is 4.2%. In the case of a relaxed-SiGe alloy, the lattice parameter can be estimated by Vegard's law, which represents the linear interpolation of the lattice constants of Si and Ge. If a thin layer of Si is pseudomorphically grown on an unstrained bulk $Si_{1-x}Ge_x$ substrate, the lattice of the epitaxially grown Si is stretched in-plane to match the lattice constant of the underlying $Si_{1-x}Ge_x$ layer. Such structure is said to be tensilely strained, referring to the in-plane strain state. Inversely, a compressively strained-SiGe film can be obtained by growing a layer with higher Ge content on a substrate with lower Ge concentration. For strain engineering, one needs to choose the appropriate type of stress to create a proper symmetry reduction and thus to produce a desirable splitting and warping.

In Si and SiGe alloys with Ge content up to 85%, the conduction band is 6 fold degenerated with a minimum in the Δ-point (Δ6 valleys), and the valence band is comprised of the degenerated heavy and light hole bands (HH and LH) with their minima located at the Γ-point. In the conduction band, the effective mass of the six-fold degenerated Δ6 valleys is anisotropic with a longitudinal mass m_l and a transverse mass m_t. The HH effective mass is higher than the LH as well as pronouncedly more anisotropic due to band warping along different crystal directions. A schematic of the conduction band energy contour and the valence band E-k dispersion for relaxed Si is shown in Fig. 6.1.

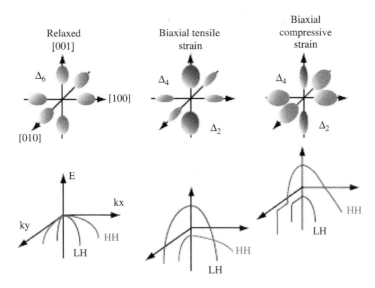

Figure 6.1 Schematics of the conduction band energy contour (top) and the valence band E-k dispersion (bottom) for relaxed Si, biaxial tensile strain, and biaxial compressive strain [1].

Silicon-germanium has emerged as a promising alternative for expanding the performance of silicon-based VLSI technology. In conventional Si/SiGe channels metal-oxide-semiconductor field-effect transistors (MOSFETs), carriers are confined in 2D potential wells as the applied transverse electric field bends the band below the Fermi level, thus accumulating carriers in the gate oxide/channel interface to form an inversion layer. In Si/SiGe MOSFETs, since the effective mass of the conduction band valleys is anisotropic, and the effective masses of HH and LH in the valence band are different, the energy of bands with lower transverse masses (in the direction of the applied electrical field) is lifted and band splitting occurs. For the p-MOSFET, the biaxial tensile strain configuration leads to serious shortcomings. Although the strain populates the lower-energy LH band that has considerably smaller effective mass, this configuration contradicts the confinement effect, where LH shifts to higher energies. As a result, the scattering rate increases and the mobility enhancement eventually disappear at high inversion charge densities. High strain levels are therefore needed to maintain the band splitting in high vertical electrical field. The impact of biaxial tensile strain on the band splitting of the conduction and valence bands of Si is schematically shown in Fig. 6.2. One of the key features of SiGe is its energy bandgap, which is lower than that of silicon which is lower than that of silicon. The bandgap of SiGe depends on the Ge mole fraction and changes from 1.17 eV for $x = 0$ to 0.78 eV for $x = 0.5$. In this range of Ge mole fractions, the dependence has an approximately linear character.

In strained-channel transistors, the final system energy band structure is determined by both, confinement quantization and strain effects. The direction and magnitude of the strain as well as the plane orientation of the applied electric field determines whether the impact of confinement and strain on mobility enhancement are additive or not.

In addition to the energy bandgap, a critical issue for device operation is the actual bandgap alignment between SiGe and Si. The two combinations of Si and SiGe that are relevant for device operation and the corresponding energy bands alignments are illustrated schematically Fig. 6.3.

Figure 6.3 shows a schematic illustration of the energy bands alignment and of the layer structure in type I and type II Si/SiGe heterostructures. In both cases the Ge mole fraction x is 0.3. The band alignment shown in the left side of Fig. 6.3 is of the type I, i.e., the SiGe bandgap lies within the Si bandgap. The bandgap difference ΔE_g is the sum of the valence band ΔE_v and of the conduction band ΔE_c offsets and is accommodated almost entirely in the valence band.

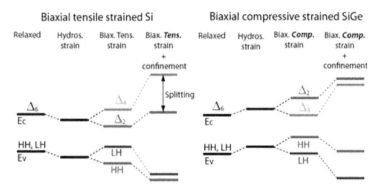

Figure 6.2 Schematics of the band splitting in biaxial tensile strained Si (left) and biaxial compressive strained SiGe (right) associated with the electrical confinement effect. Biaxial tensile strain reduces interband phonon scattering of electrons, while biaxial compressive strain enhances hole mobilities.

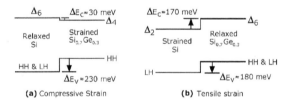

Figure 6.3 Type I and type II band offsets in a strained-SiGe layer.

The large valence band discontinuity and its monotonic increase with the Ge mole fraction are two of the most attractive features of the type I Si/SiGe alignment that can be used to control hole confinement in field-effect transistors and hole injection and electron transport in heterostructure bipolar transistors. Type I Si/SiGe/Si p-channel MOSFETs, which are relatively easy to fabricate in a VLSI process have exhibited hole mobilities higher than in Si devices. However, practical implementations of large Ge mole fractions in type I p-MOSFETs have been precluded by the necessity of complying with the Matthews-Blakeslee critical layer thickness, by the rectangular or trapezoidal shape of the Ge profile in the channel, and by interface scattering at very abrupt Si/SiGe interfaces.

The band diagram is significantly different for the second case, that of strained silicon on a relaxed-SiGe substrate. This band alignment, in which the SiGe bandgap lies partially above the silicon bandgap, is defined as type II. The type II energy bands alignment scheme allows for the separate confinement of electrons and holes at the heterojunction, that is holes in the SiGe layer and electrons in the strained-Si layer. This property of the type II system is used in both n- and p-MOSFETs. Due to the negligible conduction band offset in the type I Si/SiGe system, there is minimal electron confinement to increase n-MOSFET speed and have therefore been attempted primarily in the type II system. High electron and hole mobilities have so far been the exclusive domain of type II Si/SiGe MOSFETs. However, these devices, which require a relaxed-SiGe substrate, which is difficult to integrate in conventional Si technology. Furthermore, the defect density in type II Si/SiGe layers is high to meet VLSI specifications.

The electronic properties and material parameter models used in simulation for relaxed SiGe and the strained Si, as discussed below, are based mainly on the work reported in the literature. The band edge offsets between the relaxed-SiGe/strained-Si layers are critical in determining the heterostructure device characteristics. The type II strained Si on the $Si_{1-x}Ge_x$ heterostructure has considerable band offsets in both the conduction and valence bands. The strain induces band splitting of both degenerate conduction and valence bands, which lowers the effective mass and reduces the inter-valley scattering and therefore enhances both hole and electron transport. The $Si_{1-x}Ge_x$ alloy has been found to have an indirect bandgap and the conduction band changes from Si-like. The measured $Si_{1-x}Ge_x$ bandgap at 4.2 K using photoluminescence measurements results in the following quadratic expression as a function of Ge content is given by

$$E_g(x, 4.2K) = 1.155 - 0.43x + 0.206x^2 \ (x < 0.85)$$

To estimate the $Si_{1-x}Ge_x$ bandgap at higher temperature, one can modify the well-known relationship of temperature on the bandgap of Si. The bandgap of strained Si as a function of the Ge content, x, in the bottom relaxed-$Si_{1-x}Ge_x$ substrate at room temperature is given by:

$$E_g(x) = 1.11 - 0.6x$$

The calculation of the heterostructure band offsets is based on the average offset of the valence band and is given by

$$\Delta E_{v,av} = (0.06x - 0.47)x$$

The valence band offset, ΔE_v, is thus determined by considering the highest valence band splitting of $\Delta E_{v,av}$. The conduction band offset, ΔE_c, is calculated by subtracting the bandgap of relaxed $Si_{1-x}Ge_x$), from the sum of the strained-Si bandgap, and the valence band offset of the heterostructure, ΔE_v which may be expressed by the following quadratic expression as a function of Ge content:

$$\Delta E_v = -0.238x + 0.03x^2$$

where the negative sign indicates that the valence band of strained Si is lower than that of relaxed $Si_{1-x}Ge_x$. The conduction band splitting of strained Si exhibits a linear relationship, $\Delta E_{c,split} = 0.67x$. A cubic fit for the conduction band offset as a function of the substrate Ge content, x, has been proposed as:

$$\Delta E_c = E_{g(SSi)} + \Delta E_v - E_{g(SiGe)} - 0.35x + 0.12x^3$$

where the negative offset denotes that the conduction band edge of strained Si is lower than that of relaxed $Si_{1-x}Ge_x$, providing electron confinement in the strained-Si layer at the strained-Si/$Si_{1-x}Ge_x$ interface. The material properties used in simulation are summarized as follows:

$$\chi_{ss} = 4.17 + 0.67x$$

$$E_{g,ss} = 1.1 - 0.4x$$

$$\varepsilon_{r,ss} = 11.8$$

$$\chi_{SiGe} = 4.17$$

$$E_{g,SiGe} = 1.1 - 0.7289x$$

$$\varepsilon_{r,SiGe} = 11.8 + 4.2x$$

where x is the Ge content in the $Si_{1-x}Ge_x$ layer, χ is the affinity, E_g is the bandgap energy, and SS denotes the strained Si.

6.1.1 Hole Mobility

Electron and hole mobility in Si is strongly dependent on the temperature, doping concentration, electric field and the alloy composition in case of strained Si. The mobility in bulk relaxed-SiGe alloy is generally lower than the bulk unstrained Si and approaches towards the mobility of pure Ge at large Ge fraction due to alloy scattering. Thus the device design is critically dependent on the information about the transport parameters of the Si-based heterolayer. State-of-the-art transport modeling in strained Si consists of full-band Monte Carlo simulations using nonlocal empirical pseudo potentials method to determine the strained Si band structure. The electron and hole mobility of a relaxed, undoped SiGe layer depend on the Ge fraction in the SiGe layer.

In case of strained-Si device simulation it is necessary to take into account both the electron and hole mobility in the strained Si and also the bottom relaxed-SiGe layer. Due to presence of the Ge atom an extra scattering mechanism called alloy scattering has to be taken into account. Theoretical calculation shows that the relaxed-SiGe alloy exhibits only a moderate mobility enhancement. Also, the measured experimental room temperature mobility values of the holes in strained-SiGe layers show much less enhancement compared to the theoretical predictions. The in-plane hole mobility in compressively strained $Si_{1-x}Ge_x$ can be modeled through the modification of the low field hole mobility in Si. This is because the fact that the holes in the buried channel do not suffer from surface scattering and hence its dependence on the vertical electric field can be neglected. The doping concentration and temperature dependent low field Arora model has been modified by the hole mobility enhancement factor, which essentially takes into account the first order effect of reduction in the in-plane hole effective mass. The mobility has been calculated for various doping concentration and compared with the theoretical results. The modified Arora mobility and alloy scattering limited mobility are combined according to the Mathiessen's rule as follows:

$$\frac{1}{\mu_{Si_{1-x}Ge_x}} = \frac{1}{\mu_{Arora}\left(1 + 4.31x - 2.28x^2\right)} + \frac{1}{\mu_{alloy}}$$

where x is the Ge fraction in the compressively strained-$Si_{1-x}Ge_x$ layer and μ_{alloy} is the alloy scattering limited mobility.

In the simulation, the low field hole mobility for $Si_{1-x}Ge_x$ has been modeled where both these effects are taken into account. The doping concentration and temperature dependent mobility model due to Arora is modified by using an analytic expression involving Ge content, x, as follows:

$$\mu(x,T,N) = \mu_{Arora}\,(T,N)\cdot(1 + 4.31x - 2.28x^2)$$

where μ_{Arora} is given by

$$\mu_{Arora} = \mu_{1p}\left(\frac{T}{300}\right)^{\alpha_p} + \frac{\mu_{2p}\left(\dfrac{T}{300}\right)^{\beta_p}}{1 + \dfrac{N}{N_{cp}}\left(\dfrac{T}{300}\right)^{\gamma_p}}$$

where $\mu_{1p} = 54.3\ cm^2/V.s$, $\mu_{2p} = 407.0\ cm^2/V.s$, $\alpha_p = -0.57$, $\beta_p = -2.23$, $\gamma_p = 2.546$, and $N_{cp} = 2.67 \times 10^{17}\ cm^{-3}$.

Mobility due to alloy scattering is given by

$$\frac{1}{\mu_{alloy}} = x(1-x)\exp(-7.86x)/124.1\quad x \le 0.2$$

$$\frac{1}{\mu_{alloy}} = \exp(-2.68x)/2150 \text{ for } 0.2 < x < 0.6$$

In the simulation, the carrier mobility enhancement factor in a strained-SiGe layer perpendicular to the <100> growth plane has been modeled by modifying the Lombardi (CVT) model. In the 2D device simulation the Poisson and Schrödinger equations have to be self-consistently solved together with the drift-diffusion equations.

6.1.2   Electron Mobility

The semiconductor–insulator interface is very important for determining the mobility and hence the performance of MOSFET devices. The effective mobility of the carrier in the channel is nearly two times (or more) lower than the bulk mobility in the semiconductor. While the Coulomb and phonon scattering mechanisms mostly determine the bulk mobility in a semiconductor, surface roughness scattering plays an important role in inversion layer mobility. The electron mobility in the strained-Si surface channel has been experimentally determined by several groups. In simulation, the enhancement factor, F, for electron mobility has been empirically modeled as

$$F = 1 + 7.3x - 22x^2 + 21x^3, \text{ for } x \leq 0.4$$

where x is the Ge fraction of the relaxed-$Si_{1-x}Ge_x$ virtual substrate, creating strain in the epitaxial Si layer. The electron mobility increases up to the 20% Ge fraction in the relaxed-SiGe virtual substrate. The electron mobility has been modeled by modifying the CVT mobility model for Si as follows:

$$\mu_{s\text{-Si}}(E_\perp, T, N_A, x) = \mu_{\text{Si.CVT}}(E_\perp, T, N_A)(F)$$

The carrier mobility $\mu_{\text{Si.CVT}}$ is approximated by

$$\frac{1}{\mu_{\text{Si.CVT}}} = \frac{1}{\mu_{ac}} + \frac{1}{\mu_b} + \frac{1}{\mu_{sr}}$$

where μ_{ac} is the carrier mobility, limited by the scattering with the surface acoustics phonons, μ_b is the carrier mobility limited by the scattering with optical intervalley phonons, and μ_{sr} is the mobility limited by surface roughness. μ_{ac} is given by

$$\mu_{ac} = \left[\frac{BT}{E_\perp} + \frac{CN^\tau}{E_\perp^{1/3}} \right] T^{-1}$$

where $E_\perp$ is the perpendicular electric field, N is the total doping concentration, and T is the temperature. μ_{sr} is given by

$$\mu_{sr} = \frac{\delta}{E_\perp^2}$$

and μ_b is given by

$$\mu_b = \mu_o \exp\left(\frac{-P_c}{N}\right) + \frac{\left[\mu_{max}\left(\frac{T}{300}\right)^{-\gamma} - \mu_o \right]}{1 + \left(\frac{N}{C_r}\right)^\alpha} - \frac{\mu_1}{1 + \left(\frac{C_s}{N}\right)^\beta}$$

Strained Si under biaxial tension has been extensively investigated recently for both n- and p-channel MOSFETs. Since strained Si provides large band offsets in both conduction and valence bands and does not suffer from alloy scattering (hence mobility degradation) as in $Si_{1-x}Ge_x$, a significant improvement, especially for electron mobility, can be achieved.

By inserting a precisely controlled SiGe layer in the vertical structure of a field-effect transistor an additional design freedom is obtained. As a consequence, the current and voltage gain, unity gain cutoff frequency, and maximum frequency of oscillation can be optimized over wider voltage and temperature ranges, as compared to the original silicon devices. Because of the possibility of controlling both the bandgap and the index of refraction, strain-induced changes must be computed for the bandgap, the effective densities-of-states, and the conduction and valence band offsets. The substrate-induced strain improves the transport properties which are due to modifications of the band structure in the form of valley and band splitting and changes of the effective mass. A modified band structure leads to a changed low-field bulk mobility, which is the most basic transport quantity.

It has been reported that the strain-induced improvement of the bulk low-field drift mobility of electrons and holes as a function of the germanium content x in the $Si_{1-x}Ge_x$ substrate, the lattice mismatch of which to silicon is the origin of the biaxial tensile strain in the silicon layer. The Ge content is a measure for the involved stress; for example, a Ge content of 10% in the SiGe substrate corresponds to a stress of 664 MPa. The reported results of the piezoresistive model are based on the piezoresistive coefficients extracted from measurements in the linear stress regime. In the case of holes, the strong increase of the drift mobility is due to a continuing decrease of the effective hole mass in the top most valence band. Consequently, the piezoresistive mobility model must be modified when used in classical device simulation for higher stress levels.

In MOSFETs, surface roughness scattering occurs in addition to phonon and impurity scattering. SPARTA simulation results for the strain-induced improvement of the long-channel effective electron and hole mobility as a function of the effective field for several substrate germanium contents in comparison with measurements. While there are some quantitative deviations, the overall agreement is satisfying. In particular, the experimental trends with respect to germanium content and effective field are reproduced. In the case of electrons, the improvement saturates above a Ge content of 20%

and the dependence on the effective field is relatively weak. The effective hole mobility enhancement continues up to Ge contents of 40%, but is strongly degraded at high effective fields. In device simulation, where the total low-field mobility is decomposed according to Mathiessen's rule, the stress dependence of bulk and surface mobility, therefore, should be modeled differently. However, for short-channel MOSFETs, in the high-field regime, the influence of stress on the high-field drift velocity must be considered.

6.2 Strained-Si Channel p-MOSFETs

To improve PMOS performance yet without decreasing its size is being explored. In the following, we simulate the electrical performance of nanoscale dual-channel heterostructure biaxial strained-Si channel PMOS and single-channel biaxial strained-Si PMOS. Simulation tools used will be Silvaco TCAD tools, which are used for the simulation of fabrication using ATHENA and electrical characteristics simulation using ATLAS software. Several performance matrices including hole mobility, threshold voltage, subthreshold swing, drain induce barrier lowering, and leakage current, will be evaluated and compared with a conventional MOSFET.

In this section, typical process steps for n- and p-MOSFET fabrication using strained Si (on relaxed SiGe) as a channel are discussed. At first, the ATHENA process tool is used to create a strained-Si channel p-MOSFET structure. Typical Si MOS process steps are used. On the basis of the default materials library data and reference design provided by ATHENA, strained-silicon layers are deposited on the relaxed-SiGe layer. Subsequently, the device structure generated by the ATHENA tool is passed on to the device simulation tool ATLAS.

To study the effects of germanium concentration in the relaxed-SiGe layer, which is used to physically strain the silicon lattice, Ge content is increased from 20 to 40%. Although increase in the Ge concentration results in higher carrier mobility, it is impractical to use more than 40% Ge in the relaxed-SiGe layer due to critical layer thickness limitation of strained Si. The design of the SiGe layer has been optimized by numerical simulations using Ge fraction dependent process and device simulation model parameters. For hole mobility that takes into account Ge fraction and doping density has been used in simulation. The process steps for generating the surface channel strained-Si channel p-MOSFET structure are performed in ATHENA. Parameters used in process simulation were chosen from Reference (see Ref. [10] of Chapter 1). Physical deposition and etch models are used to provide realistic topography. The results of the process/device simulation are discussed below. This example demonstrates the interface from process to device simulation for a strained-Si p-MOSFET device.

The main parameters used in process simulation were obtained from the Silvaco ATHENA tool. Modification in process steps and optimization were carried out to obtain different devices and enhance the device performance. The gate oxide thickness, the gate length, and the Ge concentration are modified to improve device performance. As the process flow for all types of devices is almost similar, the main process steps are shown for the devices and only the differences are highlighted. The main process steps used in simulation for single-channel SiGe p-MOSFETs are as follows:

- Growth of Si epitaxial layer
- Pad oxidation (local oxidation of silicon [LOCOS] process)
- Deposition and patterning nitride
- Growth of oxide (LOCOS)
- Etching of nitride and oxide
- 8 nm Si layer deposition
- 10 nm Si layer deposition
- 15 nm SiGe layer deposition
- 7 nm strained-Si layer deposition
- Photoresist deposition for masking
- Etching of Si, SiGe, and SiO_2 layers
- 5 nm gate oxide deposition
- Poly-Si gate formation
- Boron implantation
- Nitride deposition
- Photoresist deposition and pattering
- Etching of nitride
- Photoresist removal
- Aluminum deposition and patterning

The initial n-type Si substrate with [100] surface orientation and phosphorus doping concentration of 2.0×10^{18} atoms/cm^3 is chosen for the PMOS. A 2.5 µm epitaxial layer of Si was grown on to the Si substrate by performing epitaxial growth at 800°C with a phosphorus doping concentration of 4.0×10^{16} cm^{-3}. Then another 0.5 µm thick epi-Si layer was grown. A thin pad oxide layer is grown by dry oxidation which is necessary to prevent the substrate from stress. Then LOCOS is used to achieve isolation between active regions and silicon nitride with 0.35 µm thickness was deposited by using chemical vapor deposition (CVD), which acts as a mask. The process simulation continued with the deposition of an 8 nm thick, Si layer which was doped with boron at a concentration of 1.0×10^{17} cm^{-3}. This was followed by deposition of a 10 nm Si layer onto the existing Si layer, doped with phosphorus at a concentration of 1.0×10^{16} cm^{-3}. The next process involved was deposition of a $Si_{1-x}Ge_x$ layer with the thickness of 15 nm and Ge content, x, of 0.3. This $Si_{1-x}Ge_x$ layer was also doped with phosphorus concentration of 1.0×10^{16} cm^{-3} and the structure of this simulation is shown in Fig. 6.4. The deposited thin layer of Si had created a strained Si, as shown in Fig. 6.5. A photoresist layer was then deposited on top of the Si cap and patterned. The process simulation progressed with the removal of the Si and $Si_{1-x}Ge_x$ layer that covered the LOCOS. To remove the Si and SiGe layer, etching process was carried out. The removal of the Si, SiO_2 and SiGe layer started with etching of the silicon layer and then followed by an etching of SiGe layer and lastly the oxide layer as shown in Fig. 6.6. An ATHENA-generated single-channel SiGe p-MOSFET device structure is shown in Fig. 6.4.

The gate oxide was deposited at a thickness of 5 nm, as presented in Fig. 6.7. The simulation continued with a 0.2 µm polysilicon deposition, doped with phosphorus at a concentration of 1.0×10^{15} cm^{-3} for minimizing the series resistance. The polysilicon was then etched leaving 0.1 µm for the transistor gate length. The structure of this polysilicon gate is shown in Fig. 6.8. Next, the boron-phosphorus silicate glass (BPSG) deposition for silicon nitride was carried out. The BPSG process had a low reflow temperature and it was used because it is able to provide a more planar surface for subsequent layers.

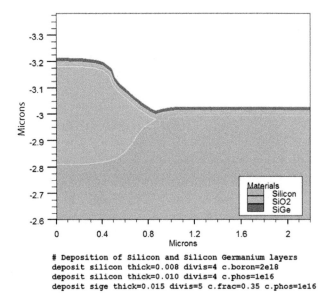

```
# Deposition of Silicon and Silicon Germanium layers
deposit silicon thick=0.008 divis=4 c.boron=2e18
deposit silicon thick=0.010 divis=4 c.phos=1e16
deposit sige thick=0.015 divis=5 c.frac=0.35 c.phos=1e16
```

Figure 6.4   The deposition of SiGe layer structure.

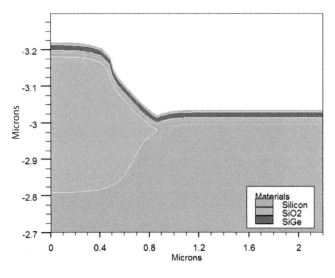

```
# Deposition of Strained Silicon layer
deposit silicon thick=0.007 divis=4 c.phos=1e16
```

Figure 6.5   The deposition of strained-Si layer.

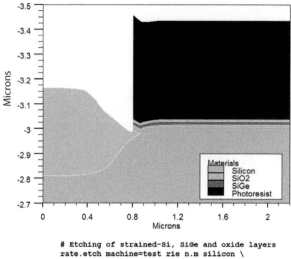

```
# Etching of strained-Si, SiGe and oxide layers
rate.etch machine=test rie n.m silicon \
isotropic=0.0 directional=20.0
rate.etch machine=test rie n.m sige \
isotropic=0.0 directional=20.0
rate.etch machine=test rie n.m oxide
isotropic=0.0 directional=3.0
```

Figure 6.6 The etching of Si, SiGe, and oxide layers.

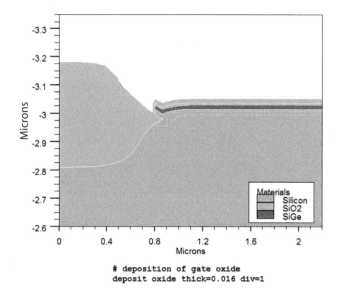

```
# deposition of gate oxide
deposit oxide thick=0.016 div=1
```

Figure 6.7 The deposition of gate oxide structure.

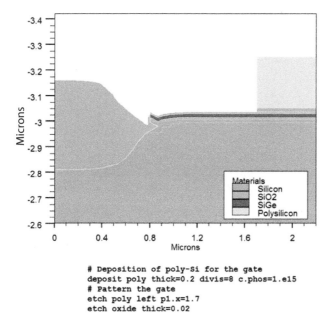

```
# Deposition of poly-Si for the gate
deposit poly thick=0.2 divis=8 c.phos=1.e15
# Pattern the gate
etch poly left p1.x=1.7
etch oxide thick=0.02
```

Figure 6.8 The patterning of polysilicon gate oxide structure.

Process simulations were continued with photoresist deposition and patterning again. This photoresist structure acts as a mask for the silicon nitride etching process. With the etching process done, the photoresist was stripped off. The deposition of the contact metal was carried out using the CVD process. Aluminum was deposited and patterned, as shown in Fig. 6.9. Next, the created device structure at the ATHENA simulator was mirrored to the right to obtain a full PMOS structure. The simulated strained PMOS with a surface channel structure from this process simulation was used later in ATLAS simulator for device simulation, to extract the electrical characteristic of this structure for analysis. Lastly, the electrodes for the PMOS structure such as gate, source, drain and substrate were labeled using the ATHENA Electrode menu. A complete structure of the 100 nm single-channel PMOS was obtained, as shown in Fig. 6.10. Various p-MOSFET device structures with different thicknesses of the strained-Si cap and different Ge contents in $Si_{1-x}Ge_x$ layer of this surface channel p-MOSFET have been generated in ATHENA for use in device simulation using ATLAS.

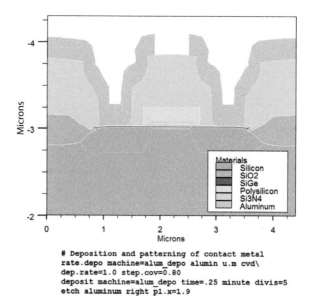

```
# Deposition and patterning of contact metal
rate.depo machine=alum_depo alumin u.m cvd\
dep.rate=1.0 step.cov=0.80
deposit machine=alum_depo time=.25 minute divis=5
etch aluminum right p1.x=1.9
```

Figure 6.9 The deposition of metal contact (aluminum) structure.

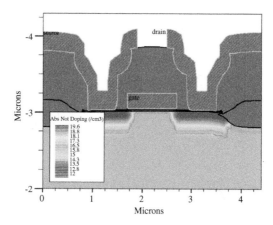

Figure 6.10 The simulated 100 nm strained-PMOS structure with net doping shown.

For this dual-channel process simulation, a 60 nm thickness of the relaxed-$Si_{1-x}Ge_x$ layer with Ge concentration of 20% was deposited on top of the Si substrate. Then a 15 nm thick, strained-$Si_{1-x}Ge_x$ layer with a Ge concentration of 30% was deposited on top of the relaxed-$Si_{1-x}Ge_x$ layer. This was followed with the deposition of 7 nm strained-Si (Si cap) layer.

Biaxial strained-Si channel p-MOSFETs are designed with the 100 nm channel length. The strained-Si p-channel devices were designed using a modified CMOS process in SUPREM4 based on the Silvaco ATHENA process simulator. Strained Si with a single channel and a dual channel are designed. Strained Si with single-channel design is created when a SiGe layer is grown on the Si substrate, followed by growing the strained-Si layer on the SiGe layer. Strained Si with dual-channel design was created with a relaxed $Si_{1-x}Ge_x$ is deposited on a silicon substrate, followed by pseudomorphically grown strained-Si/SiGe channel layers to the relaxed-$Si_{1-x}Ge_x$ layer. A control device in which Si is unstrained was also designed using the ATHENA process simulator. Main process steps for the fabrication of heterostructure MOSFETs are listed. The process specifications are obtained from reported experimental devices for designing the process flow.

The following main process steps were used in ATHENA for dual-channel p-MOSFET device structure:

- Growth of Si epitaxial layer
- Pad oxidation (LOCOS process)
- Deposition and patterning nitride
- Growth of oxide (LOCOS)
- Etching of nitride and oxide
- 8 nm Si layer deposition
- 10 nm Si layer deposition
- 60 nm SiGe layer deposition
- 15 nm SiGe layer deposition
- 7 nm strained-Si layer deposition
- Photoresist deposition for masking
- Etching of Si, SiGe, and SiO_2 layers
- 5 nm gate oxide deposition
- Poly-Si gate formation
- Boron implantation
- Nitride deposition
- Photoresist deposition and pattering
- Etching of nitride
- Photoresist removal
- Aluminum deposition and patterning

The ATHENA-generated final dual-channel p-MOSFET device structure is shown in Fig. 6.11 The doping profile is shown in Fig. 6.12.

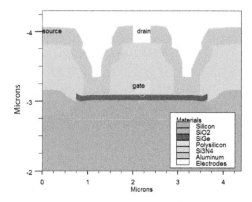

Figure 6.11 The process-simulated buried (dual)-channel strained-Si p-MOSFET structure.

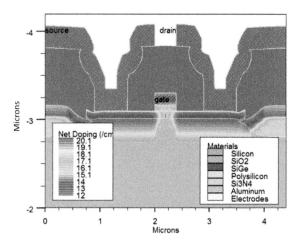

Figure 6.12 The simulated 100 nm buried-channel strained-PMOS structure showing the doping profile.

The simulated strained PMOS with a dual-channel structure from the process simulation was used in ATLAS for device simulation. This is a necessary step in order to extract the electrical characteristics of the structures. The important material parameters needed for device simulation include:

- Valence and conduction band offsets
- Electron and hole mobilities
- Bandgap and bandgap narrowing
- Effective mass for electron and holes
- Density of states for conduction and valence bands
- Dielectric constant
- Saturation velocities of electrons and holes
- Minority carrier lifetime
- Recombination, including Auger and radiative
- Coefficient for impact ionization

Table 6.1   Material properties used in simulation

Material property	Si	Ge	SiGe
Lattice constant	5.431 Å 300 K	5.658 Å 300 K	$Si_{1-x}Ge_x$ $(5.431 + 0.20x + 0.027x^2)$ Å 300 K
Dielectric constant	11.7 300 K	16.2 300 K	$Si_{1-x}Ge_x$ $11.7 + 4.5x$ 300 K
Effective electron mass (longitudinal)	$0.98m_0$ 300 K	$1.6m_0$ 300 K	$Si_{1-x}Ge_x$ $0.92m_0$ 300 K, $x < 0.85$ $0.159m_0$ 300 K, $x > 0.85$
Effective electron mass (transverse)	$0.19m_0$ 300 K	$0.08m_0$ 300 K	$Si_{1-x}Ge_x$ $0.19m_0$ 300 K, $x < 0.85$ $0.08m_0$ 300 K, $x > 0.8$
Density of states			$Si_{1-x}Ge_x$ $1.06m_0$ 300 K, $x < 0.85$ $1.55m_0$ 300K, $x > 0.85$
Effective hole masses (heavy) m_{hh}	$0.537m_0$ 4.2 K	$0.33m_0$	
Effective hole masses (light) m_{lh}	$0.153m_0$ 300 K	$0.0430m_0$ 300 K	
Effective hole masses (spin–orbit–split) m_{so}	$0.234m_0$ 300 K	$0.095m_0$ 300 K	$Si_{1-x}Ge_x$ $(0.23 - 0.135x)$ m_0 300 K

Source: Ref. [2]

Besides that, a suitable model, such as the bandgap-narrowing model, the impact ionization model, etc., has been chosen in order to run this simulation. The net doping profile (vertical cutline) for a dual-channel strained PMOS is shown in Fig. 6.13. Other device information such as on the Ge profile, band diagram, hole concentration, electric field, output characteristics, and *C–V* characteristics is shown in Figs. 6.14 to 6.20, respectively.

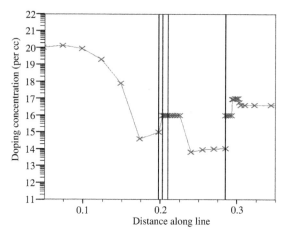

Figure 6.13 Net doping profile (vertical cutline) for dual-channel strained PMOS.

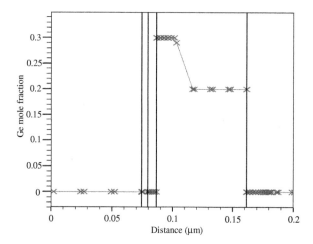

Figure 6.14 Ge content versus depth in dual-channel strained PMOS.

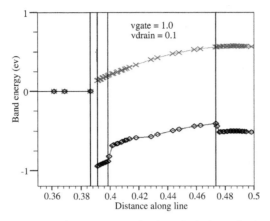

Figure 6.15 The band diagram versus depth in dual-channel strained PMOS.

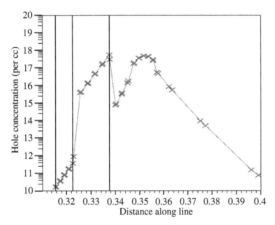

Figure 6.16 The hole concentration versus depth in dual-channel strained PMOS.

The family of I_d – V_d (output) characteristics was generated at different gate voltages (–0.8 V, –1.0 V, –1.2 V, and –1.4 V) by obtaining solutions at each of the stepped bias and was followed with a drain voltage sweep (0 V to –3.0 V) with a step voltage of –0.3 V for each bias point. Figure 6.18 shows the simulated output characteristics for strained-Si device. Drain current versus gate voltage, I_d – V_g, characteristics were generated by obtaining solutions at each of the stepped bias and were followed with a gate voltage sweep for each bias point. The drain voltage was biased to –0.1 V, –0.2 V, and –1.0

V, while the gate voltage was ramped from 1.0 V to –4.0 V with a voltage step of –0.2 V.

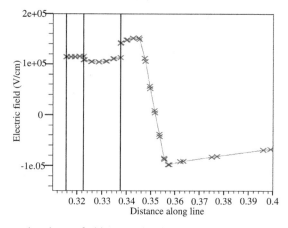

Figure 6.17 The electric field versus depth in dual-channel strained PMOS.

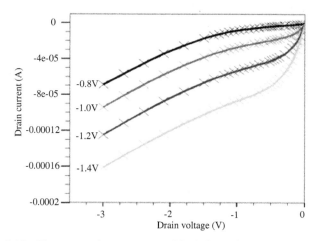

Figure 6.18 The output characteristics of dual-channel strained PMOS.

Gate capacitance versus gate voltage characteristics were obtained by ramping the gate voltage from –4 V to 4 V with a 0.2 V voltage step at 1 kHz using the quasi-static statement. Figure 6.20 shows the simulated characteristics for strained-Si devices. The strained-Si devices show a plateau in the accumulation region due to hole confinement.

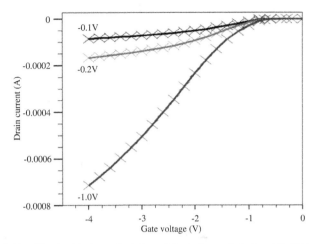

Figure 6.19 The $I_d - V_g$ characteristics of dual-channel strained PMOS.

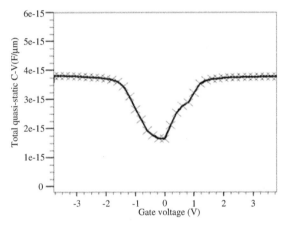

Figure 6.20 The quasi-static capacitance versus gate voltage characteristics of dual-channel strained PMOS.

6.3 Summary

With its high speed, low voltage, and low power features, Si/SiGe technology has the potential to integrate all these functions on a single chip. In this chapter, some important device simulation results for surface and buried-channel strained-Si p-MOSFETs are presented. This study provides a guideline for the design of strained-silicon channel p-MOSFETs. Simulations provide more realistic results and will allow researchers to gain a better understanding of the effects of different device parameters on the overall device performance without fabrication. Strained-Si channel p-MOSFETs require the more attention to enhance the device performance.

References

1. Minamisawa, R. A. (2011). *Physical Studies of Strained Si/SiGe Heterostructures: From Virtual Substrates to Nano Devices*, PhD thesis, Informatik und Naturwissenschaften der RWTH Aachen, University zur Erlangen.

2. Ioffe Physico-Technical Institute. *New Semiconductor Materials Characteristics and Properties*, http://www.ioffe.ru/.

Chapter 7

FinFETs

The most revolutionary MOSFET architecture is a double-gate MOSFET which was first proposed in the early 1980s and represented by FinFET in late 1990s. It has been recently regarded as the ultimately scalable MOSFET. In the area of multigate transistors, double-gate finFETs are considered a serious contender for channel scaling because of their quasi-planar structure and the compatibility with CMOS. The salient features of the double-gate MOSFET include (1) reduced 2D short-channel effects, leading to a shorter allowable channel length compared to bulk MOSFET; (2) a sharper subthreshold slope, which allows for a larger gate overdrive for the same power supply and the same off-current; and (3) better carrier transport behavior as the channel doping can be reduced.

7.1 Basics of FinFETs

Fin-shaped field-effect transistors (finFETs) provide excellent electrostatics and short-channel effect (SCE) control, and can tolerate low channel doping which reduces the dopant-induced variability. A nonclassical structure, such as a finFET, is schematically shown in Fig. 7.1. The silicon fin is surrounded by two sidewall gates and optionally by a top one, thus providing a better short-channel control. Charge transport is therefore a real 3D phenomenon,

Introducing Technology Computer-Aided Design (TCAD): Fundamentals, Simulations, and Applications
C. K. Maiti
Copyright © 2017 Pan Stanford Publishing Pte. Ltd.
ISBN 978-981-4745-51-2 (Hardcover), 978-1-315-36450-6 (eBook)
www.panstanford.com

composed of two current flows parallel to the fin sidewalls and, optionally, an additional third one at the fin top. Compared with bulk finFETs, silicon-on-insulator (SOI) finFETs exhibit many compelling advantages:

- Shallower junction depth (lower junction capacitance) due to the natural barrier (BOX) against dopants diffusion
- No punch-through due to the thin film and BOX
- Higher mobility and reduced threshold voltage mismatch due to low-doped channel
- Better control of SCEs
- Easier mobility boosters such as strained SOI and SiGe

However, 3D structures face some challenges:

- For further scaling, more advanced photolithography is needed to fabricate narrower fins.
- The enhanced quantum confinement in an extremely narrow fin can cause mobility degradation.
- The coupling effect between the multiple gates is amplified in narrow finFETs.
- The corner effect amplifies the local electric field, so optimized design is needed.

At the early stage of novel technology development, the application of technology computer-aided design (TCAD) can give a physical insight and guideline for process optimization. The finFET is one such emerging device which is considered to be a suitable successor to the conventional metal-oxide-semiconductor field-effect transistor (MOSFET) winning over many of the hurdles mentioned above. It is so called because the thin channel region (body) stands vertically like the fin of a fish between the source and drain regions. The gate wraps around the body from three sides, and this is responsible for higher gate channel control and therefore reduced SCE.

After the first introduction of the 3D finFET architecture at the 22 nm node by Intel in 2011, it is now considered as a major candidate for 14 nm node mainstream technology. FinFETs provide excellent electrostatics and SCE control, and can tolerate low channel doping which reduces the dopant-induced variability. Recently, there is strong interest in finFET technology on bulk for lower cost and good compatibility with planar complementary metal-oxide-semiconductor (CMOS). Intel's 22 nm CMOS node is the first commercially available bulk finFET technology and opens a new era of 3D CMOS for the low-power applications.

Table 7.1 Operation characteristics and design considerations for various multiple-gate devices

DG design	Current direction	Electric field from gate	Design considerations
Planar	$\parallel$ to substrate	$\perp$ to substrate	Precise control of silicon thickness and bottom gate dimension, gate alignment
Fin	$\parallel$ to substrate	$\parallel$ to substrate	High aspect ratio and short pitch fin definition, nonplanar gate stack patterning
Vertical	$\perp$ to substrate	$\parallel$ to substrate	S/D doping and contacts, gate isolation
Trigate	$\parallel$ to substrate	$\parallel$ and $\perp$ to substrate	Active area hard mask removal and surface prep, layout efficiency
GAA	$\parallel$ or $\perp$ to substrate	variable	Access for gate stack deposition and etch, active area dimension uniformity

Source: Ref. [1]

Multiple-gate field-effect transistors (mugFETs) show great promise as alternative to planar CMOS technologies below gate lengths below 15 nm, as evidenced by endorsement from the International Technology Roadmap for Semiconductors. Several designs depicted in Fig. 7.1 have been proposed, including planar, vertical, fin, trigate, and gate all around (GAA), that all make use of enhanced gate control due to the action of multiple electrodes surrounding the channel. The key operation differences and technological issues for each device type are summarized in Table 7.1.

Multiple-gate devices do not require any channel doping to reduce the depletion width (and subsequent decrease in gate length) as this parameter is defined primarily by the active volume. Moreover, the S/D junction depth is similarly defined by the silicon body thickness and shifts the onus from developing complicated Halo/LDD doping schemes to ensure uniform body-doping profiles with appropriate activation anneals to control diffusion near the gate edge. While the manufacturability of mugFETs remains challenging due to their 3D structure, most utilize self-aligned gate electrodes amenable to conventional planar CMOS process modules.

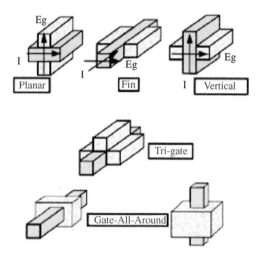

Figure 7.1 Various mugFET structures [1].

Of the varieties discussed above, the finFET and trigate are reported to be the most manufacturable. However, a 3D scaling paradigm that forms channel width perpendicular to the substrate requires building mugFETs as tall as possible (>50 nm) to maximize the channel width per die area benefit. If the trigate device height is increased accordingly, the body thickness to height ratio must migrate away from 1:1 toward ratios typically seen on finFETs to maintain channel control. In this scenario, the top gate is no longer playing an important role and the device behaves as a dual-gate (DG) field-effect transistor (FET). Hence, the best channel control, manufacturability, scalability, and channel width per die area can be achieved with the finFET structure.

Numerous technical reports and academic papers show that scaled finFET manifest itself with a higher I_{on}/I_{off} ratio, a lower subthreshold slope SS and an improved drain-induced barrier lowering (DIBL) by the enhanced gate controllability compared to conventional planar counterpart. The ultimate electrical control of the conductive channel is expected to be realized by the GAA (nanowire) transistors. The fundamental idea of the GAA transistors is similar to that of the multigate transistors, except that the conductive channel is thoroughly surrounded by the gate material. The cross-sectional views of the conductive channels of the double-gate finFETs, trigate transistors, and GAA transistors are drawn together in Fig. 7.2 to compare. The conductive channel material is a bunch of nanowires, including Si nanowires and III–V nanowires. The outstanding controllability of the conductive channel is expected to introduce tremendous improvement of the performance, and maximize the density of the integrated circuits. Recent progresses at Intel's 22 nm node not only shows the phasing out of the planar CMOS and the dawn of a 3D CMOS (finFET) era, but also the start of the low power for mobile applications as a new driving force of device scaling and extending Moore's law.

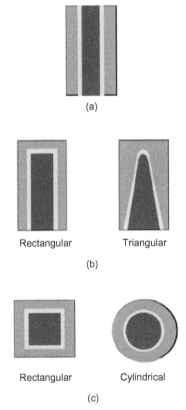

(a)

Rectangular   Triangular

(b)

Rectangular   Cylindrical

(c)

Figure 7.2 Cross-sectional views of the conductive channels of (a) double-gate finFETs, (b) trigate transistors, and (c) gate-all-around transistors [2].

In this chapter we shall address briefly the manufacturing issues of advanced CMOS process and present process solutions to the new and novel set of problems faced in 28 nm planar and 22/14 nm finFET technology. Key processing details are presented, as well as previews of the main manufacturing issues at the 10 nm node. FinFET-based CMOS and memory cells are very promising for sub-32 nm technology node. In order to familiarize the reader with the established methods of finFET device fabrication and to illustrate the compatibility with current planar CMOS processing techniques, a simplified process flow is discussed in the following section.

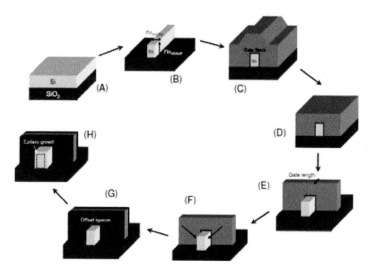

Figure 7.3 Generic multigate device fabrication flow [3]. (A) SOI wafer with predefined silicon thickness (fin height). (B) Lithography for fins followed by plasma etching. (C) Growth of gate dielectric and deposition of metal gate. (D) Gate stack deposition and planarization. (E) Gate patterning and gate etching. (F) Low-energy implant for source/drain regions. (G) Epitaxial source/drain region and nitride spacers along sidewalls of gate/fin. (H) Spacer removal and extended source/drain region formation.

The finFET process flow starts the fin formation similarly as the formation of an active area (in planar CMOS) and is followed by STI gap fill and planarization and oxide recessing to reveal the fins. The Si surface on fins appears different than in bulk. Then the rest of flow proceeds to similar steps (e.g., well, gate, epi-S/D, etc.) as the planar

CMOS with gate-last high-k, and metal gate (HKMG) flow. As a result of the fin shape, the low doping in channel is preferred for minimizing threshold voltage variations. Vertical multigate transistors such as finFETs, Ω-FETs, and trigate FETs have self-aligned gates, which is a major advantage from a fabrication perspective. The major fabrication steps for FEOL processing of generic SOI finFETs is shown in Fig. 7.3.

7.1.1 Stress-Enhanced Mobility in Embedded SiGe p-MOSFETs

From 90 nm technology node onward, strained silicon is in use for improving transistor performance in scaled devices. Mobility gains as high as 50% and drive current enhancements of 25% have been demonstrated in 45 nm gate length devices. Stress engineering has clearly shown that due to stress generated by the stressors mobility enhancement takes place in individual n-finFETs. However, when different devices (e.g., p- and n-FETs) are placed in close proximity (e.g., in a cell) with both n- and p-liners, noticeably different stresses and mobility enhancements can be obtained. This happens because of the local nature of S/D stressors and the interaction effect for different S/D stressor configurations or for denser cell layouts. 3D process and stress simulation is useful for the detailed performance analysis and optimization of various stress engineering schemes by varying geometry of stressors, substrate orientation, and material composition of each device, as well as by changing location and density of the individual devices inside the cell layout. Mechanical stress sources used in state-of-the-art CMOS technologies are shown in Fig. 7.4.

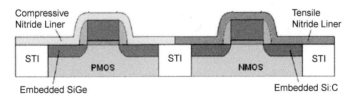

Figure 7.4 Mechanical stress sources in a CMOS transistor [4].

7.2   Stress-Engineered FinFETs

Strain was implemented intentionally as a mobility booster in devices that went into production in 2004. Besides the salient effects on the charge carrier mobilities, stress affects the width of the bandgap and has widespread direct and indirect effects on the defects in semiconductors. As the device width shrinks, 3D simulation might first be routinely employed in mechanical stress investigations. It is therefore our aim to investigate by full 3D stress, process and device simulation. Stress has significant influences on the electrical performance and hence where it should be optimized.

It has been shown that a simple piezoresistance mobility model can describe the stress impact on transistor performance with good accuracy. The piezoresistance model provides accurate stress-dependent mobility values (within about 20%) at stress levels below 1 GPa. At higher stress levels, holes exhibit super linear mobility gain with increasing stress, whereas electron mobility gain with stress becomes sublinear and eventually saturates. The following equations are used for the calculation of n- and p-mobility enhancement factors for (100)/<100> and (110)/<100> crystallographic orientations in silicon:

$$\mu_n{}^{100} = 1.0 - (-1.022\sigma_{xx} + 0.534\sigma_{yy} + 0.534\sigma_{zz})$$
$$\mu_p{}^{100} = 1.0 - (0.066\sigma_{xx} - 0.011\sigma_{yy} - 0.011\sigma_{zz})$$
$$\mu_n{}^{110} = 1.0 - (-0.311\sigma_{xx} - 0.175\sigma_{yy} + 0.534\sigma_{zz})$$
$$\mu_p{}^{110} = 1.0 - (0.718\sigma_{xx} - 0.663\sigma_{yy} - 0.011\sigma_{zz})$$

In these equations the stresses are in units of GPa and coefficients of piezoresistivity are in 1/GPa. Considering that the piezoresistance model is a linear superposition of the contributions from the principal stress components, one can focus on optimizing the transistor to achieve the optimal stress pattern in the channel. However, the most beneficial stress components are different for NMOS and PMOS transistors. The NMOS benefits most from compressive vertical stress and also somewhat from tensile stress in the lateral directions, whereas the PMOS benefits most from the compressive stress in the direction along the channel. A tensile CESL cap layer simultaneously introduces compressive vertical and tensile lateral stresses that enhance electron mobility. It is important to simulate stresses not just in an individual device but in the whole cell. In the following, a combination of 3D process simulator VictoryCell and 3D stress simulator VictoryStress will be employed for stress analysis in cell structure. The inverter cell has two p-finFETs located parallel to each other and one n-finFET. We use the 3D simulator VictoryStress to analyze stress effects on carrier mobilities of individual n-finFET and p-finFET devices.

The most common method of introducing desirable stresses in the transistor channel is the deposition of high-tensile or high-compressive films of nitride-type materials. A tensile film (when intrinsic stress value is positive) improves drive current in n-MOSFETs due to enhancement of electron mobility, whereas a compressive film (when intrinsic stress value is negative) enhances the hole mobility for p-MOSFETs. The following simulation example shows stress simulation in a single n-finFET. We start with process simulation, which includes the following steps:

- Definition of geometrical domain and region limits for subsequent consistent mesh formation
- VictoryCell process steps
- Cartesian mesh formation and saving the structure for stress calculation
- Stress calculation
- Visualization and analysis of simulation results, extraction of average stresses and mobility enhancement factors

7.2.1 VictoryCell Process Steps

The common method of stress engineering is by the introduction of so-called source/drain stressors (SiGe or SiC). The structure in this example allows to have such stressor and is called "nstressor." The thickness of the stress liner over the gate is very important as it determines the amount and the stress distribution inside the fin. The finFET contains vias, which are placed over the metal contacts (aluminum). For simplicity the length of each via (for source and for drain) was chosen the same as the length of stressors (0.3 μm). Two important considerations to form an optimal mesh for Manhattan-type structures are that (a) a reasonable number of mesh points (spacing) should be generated across areas of interest such as gate and fin and that (b) it is preferable that in each direction a mesh line coincide with the material region boundary. We explore several technology options for the enhancement of electron and hole mobility in complementary MOSFETs, focusing on strain engineering using lattice-mismatched source/drain (S/D) materials. Silicon-carbon $(Si_{1-y}C_y)$ and silicon-germanium $(Si_{1-x}Ge_x)$ have lattice constants different from that of the Si channel. When $Si_{1-y}C_y$ or $Si_{1-x}Ge_x$ is embedded in the transistor S/D region, lateral tensile or compressive strain is induced in the adjacent Si channel, leading to improvement in the electron or hole mobility, respectively.

For the simulation of an n-finFET the following parameters were used:

gate_length = 0.06

gate_height = 0.05

stress_liner_height = 0.05

fin_width = 0.03

fin_height = 0.06

fin_length = 0.8 + $gate_length

oxide_box_height = 0.6

oxide_top_height = 0.15

spacer_thick = 0.005

met_height = 0.005

stress_value = 1.0×10^{10}

The following process steps were used for the creation of finFET structure using VictoryCell:

- Formation of the n-fin
- Formation of two p-fins
- Source side local stressor (nfin)
- Drain side local stressor (nfin)
- Source side local stressor (pfin)
- Drain side local stressor (pfin)
- Oxide spacer deposition
- Gate formation
- Deposition of metal contact layer

- Deposition of stress liner (n-liner)
- Deposition of stress liner (p-liner)
- Deposition of the top oxide layer
- Etching of top oxide and stress liner for the via formation
- Deposition of metal (aluminum) for formation of via

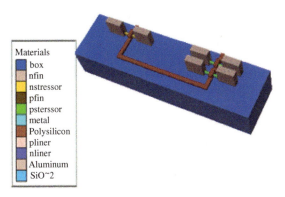

Materials
- box
- nfin
- nstressor
- pfin
- psterssor
- metal
- Polysilicon
- pliner
- nliner
- Aluminum
- SiO~2

Figure 7.5 3D view of an inverter cell having two p-finFETs located parallel to each other and one n-finFET. Only a BOX substrate and an active area of devices are shown.

By default the absolute value of the initial intrinsic stress is 1 GPa; for the n-FET it is positive; for the p-FET it is negative. The following material parameters are used for user-defined materials in the structure:

stress_value = 1.0×10^{10}

material silicon nitride ("nliner") young.m = 3.89×10^{12}

material Silicon nitride ("nliner") poiss.r = 0.3

material silicon nitride ("pliner") young.m = 3.89×10^{12}

material silicon nitride ("pliner") poiss.r = 0.3

material oxide ("box") young.m = 6.6×10^{12}

material oxide ("box") poiss.r = 0.2

material silicon ("nfin") young.m = 1.67×10^{12}

material silicon ("nfin") poiss.r = 0.28

material silicon ("pfin") young.m = 1.67×10^{12}

material silicon ("pfin") poiss.r = 0.28

material silicon ("nstressor") young.m = 1.67×10^{12}

material silicon ("nstressor") poiss.r = 0.28

material silicon ("pstressor") young.m = 1.67×10^{12}

material silicon ("pstressor") poiss.r = 0.28

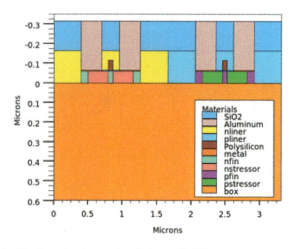

Figure 7.6 2D view showing materials in a p-finFET.

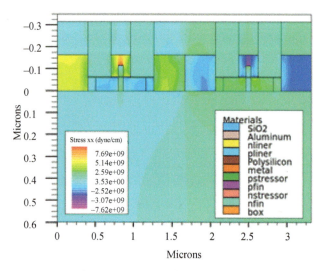

Figure 7.7 2D view showing stress distribution in a p-finFET (which is parallel to an n-finFET) first sidewall of the p-fin under the gate.

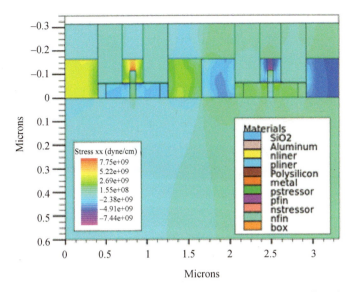

Figure 7.8 2D sidewall S_{xx} stress distribution through the center of a p-finFET.

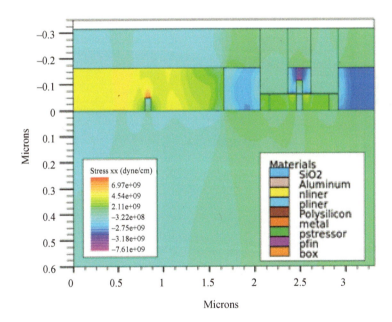

Figure 7.9 2D next sidewall S_{xx} stress distribution through the center of a p-finFET.

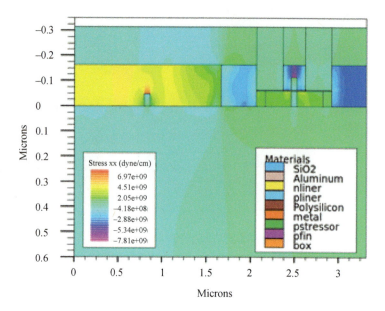

Figure 7.10 2D top-wall stress distribution through the center of a p-finFET.

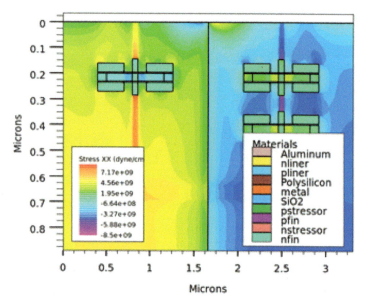

Figure 7.11 2D top-side S_{xx} stress distribution through the center of an n-finFET.

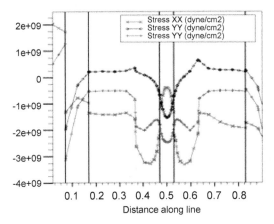

Figure 7.12 1D stress distribution under the gate used for calculation of average stress.

7.2.2 Visualization and Analysis of Simulation Results, Extraction of Average Stresses, and Mobility Enhancement Factors

Since the simulation results are in 3D, the only convenient way of analyzing them is to extract figures of merit from inside using cut planes within the 3D structure. The five planes selected are as follows: The first four planes are parallel to the fin length; two of those planes are along sidewalls of the first n-finFET and first p-finFET; the next two planes are along the sidewall of other second p-finFET; and the final plane (the fifth) is along the top of the fins. These 2D planes are generated using the "-cut" option of Tonyplot3D. The exact position of the cut could be set manually using the cut plane setting in the cut plane view menu of Tonyplot3D. The generated cut plane structures with 2D S_{xx} distribution and 1D stress distribution are shown using TonyPlot in Figs. 7.6 through 7.12.

Average stress calculations were performed using "factor" (100 in this case), which is required for obtaining integrated stresses in GPa. The S_{xx}, S_{yy}, and S_{zz} components in the sidewalls and top side of the fin under the gate were computed by integration of corresponding stress distributions along the cut 5 nm below the fin-oxide boundary under the gate.

Mobility enhancement calculations were performed using the piezorestivity coefficients for (100)/<100> n-fin orientation (in cm^2/GPa) values:

nx1 = −1.022

ny1 = 0.534

nz1 = 0.534

The piezorestivity coefficients for (110)/<100> n-fin orientation (in cm^2/GPa) values used are:

nx2 = −0.311

ny2 = −0.175

nz2 = 0.534

Mobility enhancement factors were calculated at the sidewall and top of the fin under the gate. Mobility enhancements for n-finFET with 2 fin orientations were found to be dependent on fin orientation, and also the enhancement factors are different because the piezorestivity coefficients are different. For the simulation of a p-finFET device having the same structure as n-finFET, the stresses can be calculated using the same procedure but with different piezorestivity coefficients.

The piezorestivity coefficients for (100)/<100> p-fin orientations (in cm^2/GPa) are:

px1 = 0.066

py1 = −0.011

pz1 = −0.011

The piezorestivity coefficients for (110)/<100> p-fin orientations (in cm^2/GPa) are:

px2 = 0.718

py2 = −0.663

pz2 = −0.011

The mobility enhancement factors for the first p-finFET (it is parallel to n-finFET): (100)/<100) are:

p_factor_side1_p1_1 = 1.00215

p_factor_side2_p1_1 = 1.00137

p_factor_top_p1_1 = 1.00181

The mobility enhancement factors of the first p-finFET (it is parallel to n-finFET): (110)/<100) are:

p_factor_side1_p1_2 = 0.910603

p_factor_side2_p1_2 = 0.899863

p_factor_top_p1_2 = 0.905628

The mobility enhancement factors of the second p-finFET: (100)/<100) are:

p_factor_side1_p2_1 = 1.00195

p_factor_side2_p2_1 = 1.00289

p_factor_top_p2_1 = 1.00248

The mobility enhancement factors of the second p-finFET: (110)/<100) are:

p_factor_side1_p2_2 = 0.902494

p_factor_side2_p2_2 = 0.91404

p_factor_top_p2_2 = 0.909027

A combination of the 3D process simulator VictoryCell and the 3D stress simulator VictoryStress has been employed for fast and accurate stress analysis of individual finFETs and also complex cell structures. It is important to simulate stresses, not just in an individual device, but also in the whole cell.

7.3 FinFET Design and Optimization

In this section a 3D finFET is simulated using the commercial package of Taurus Process and Taurus Device simulators along with Taurus Layout and Taurus Visual from Synopsys. Design issues for a state-of-the-art finFET starting from process to device simulation are employed to vary fin width (W_{fin}), fin height (H_{fin}), gate oxide thickness, buried oxide thickness, and various process parameters (such as channel implantation dose and energy) of a trigate SOI finFET device. We study the DC an AC behavior, including the DIBL, subthreshold swing, I_{off}/I_{on} behavior, and cutoff frequency.

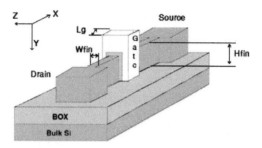

Figure 7.13 3D finFET structure used in simulation using Taurus Process and Taurus Device.

Since the finFET structure is 3D and complex channel profile is required to adjust threshold voltage, it cannot be simulated using the conventional 1D or 2D simulators. The 3D simulation requires dense meshing/gridding and thus takes large time and computer memory resources for process simulation. Only one quarter of the device was simulated first and at the end of the process simulation, the structure was reflected twice to make the complete device. The initial grid definition is by default taken from the mask layout file at the beginning of process simulation. The mask is held in the *X-Z* plane and the growth direction is in the negative *Y* axis. The process steps included are initialization of the silicon substrate, deposition of a 100 nm buried oxide layer, deposition of a 60 nm hard mask (TEOS), deposition of a 45 nm Si layer, patterning of the silicon fin using dry etching, implantation of boron in the channel with (energy = 20 keV, dose = 4.5×10^{13} cm^{-2}) , sacrificial oxidation (optional), 3 nm gate oxide , 100 nm gate polysilicon with phosphorous doping of 4×10^{20} cm^{-3}, deposition of a 65 nm TEOS hard mask, patterning of the gate, a nitride spacer of 65 nm, selective silicon epitaxy for raised source/drain of 60 nm, implantation of As in the source/drain region (energy = 50 keV, dose = 3×10^{15} cm^{-2}), and final anneal. For computational efficiency models the PDFermi (equilibrium point defect concentrations) and dual Pearson models were used for diffusion and ion implantation with their default model parameters. All necessary thermal steps included in the process simulation. All the devices were process simulated in the traditional (100) plane.

The device simulation is performed in Taurus Device simulator using the structure file created by Taurus Process. The drift-diffusion model was used in the device simulation along with the quantum-mechanical model *modified local density approximation* (MLDA), which is capable of calculating the confined carrier distribution near the Si–SiO$_2$ interface for both inversion and accumulation and simultaneously for hole and electrons. The MLDA model is capable of performing DC, transient, and AC analysis. The bandgap narrowing effect due to heavy doping is simulated using the Jain and Roulston model. Both models were used with their default model parameters. The low-field mobility models include impurity and temperature-dependent Aurora mobility model and the Lombardi surface scattering model with their default model parameters. The Caughey–Thomas mobility model was used for the high-field (along the direction of the current flow) regime. Due to the ultrashort gate length the quasi-ballistic effects were included by using a gate length–dependent velocity model in the high-field regime. The first the process flow was simulated for a finFET with W_{fin} = 20 nm, H_{fin} = 45 nm, and L_g = 20 nm and gate oxide thickness of 3 nm and the device simulation was done for calibration with the experimental data, as shown in Fig. 7.14.

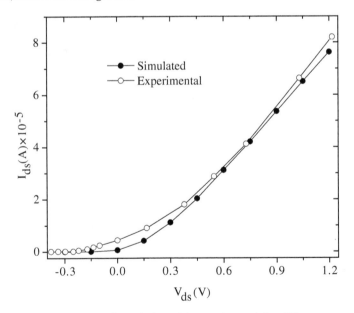

Figure 7.14 Validation of simulation with experimental data [5].

7.3.1 Simulation Setup

A schematic of a trigate SOI finFET is shown in Fig. 7.13. The axes shown in the above figure corresponds to the axes definition of Taurus Process. The axes in the schematic refer to that of the process simulated device. The fin height (H_{fin}), the fin width (W_{fin}), and the gate length (L_g) are as shown in the figure. The gate wraps from three sides and the gate oxide thickness is the same on the three sides. As seen from the figure the three gates are connected together and the same bias is applied at three sides of the gate. If a thick spacer oxide as compared to the sidewall oxides is at the place of the top gate oxide, then even though the gate wraps around from three sides, the finFET will be termed "double-gate finFET." The gate width (W_g) in the case of a trigate finFET is given by $2 \times H_{fim} + W_{fim}$, and for a double-gate device it is $2 \times H_{fim}$. With the same process flow, the fin width was varied from 10 to 20 nm with a step of 5 nm. For each fin width the fin height was varied from 40 to 70 nm with a step of 10 nm. In all cases the gate length was kept 25 nm and the gate oxide thickness was kept at 1.1 nm. Thus the aspect ratio for the devices was varied from 2 to 7.

The transfer characteristics of the devices are shown in Fig. 7.15a–d. The variation of the drain currents with variation of the fin width for fixed fin heights is shown in Fig. 7.15e,f for a fixed drain voltage and a fixed gate voltage. It is clear from the graphs that the drain current falls with the increase in the fin height for a fixed fin width, while it increases with fin width for a fixed fin height. The gate width is different for different device geometries, so the drain current was normalized for the plot. From Fig. 7.15g it can be seen that the drain current drops with increases in the aspect ratio.

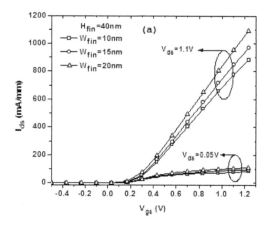

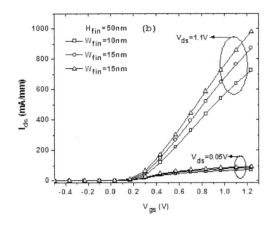

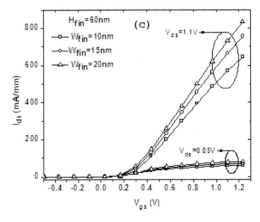

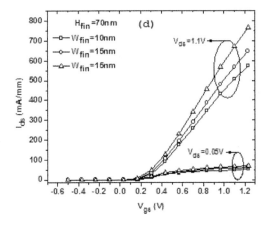

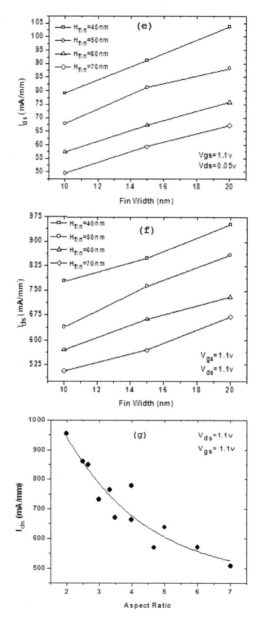

Figure 7.15 (a–d) Transfer characteristics of a finFET with fixed H_{fin} and varying W_{fin}, (e) comparison of normalized drain currents at $V_g = 1.1$ V and $V_d = 0.05$ V at fixed H_{fin} and varying W_{fin}, (f) comparison of normalized drain currents at $V_g = 1.1$ V and $V_d = 1.1$ V at fixed H_{fin} and varying W_{fin}, and (g) drain current versus aspect ratio plot at $V_g = V_d = 1.1$ V [5].

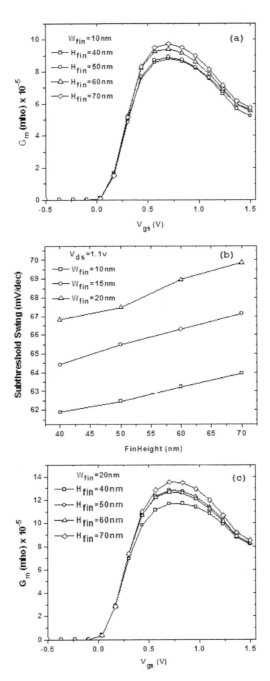

Figure 7.16 (a–c) Transconductance versus gate voltage plots at drain voltage of 1.1 V [5].

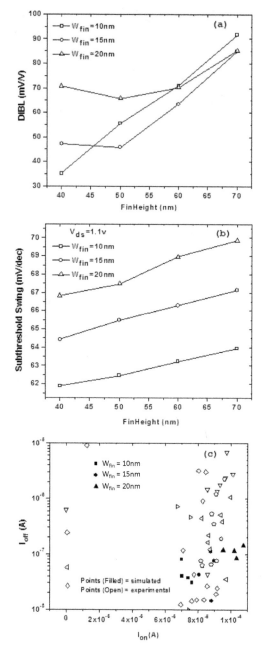

Figure 7.17 (a) Variation of DIBL with fin height for a fixed fin width, (b) variation of subthreshold swing with fin height for fixed fin widths, and (c) plot of I_{off} versus I_{on} [5].

Figure 7.16a–c shows the transconductance versus gate voltage plots at drain voltage of 1.1V. From Fig. 7.17a,b, it is seen that as the fin height is increased keeping the fin width at a fixed value both the DIBL and subthreshold swings increase. So as the fin height is increased the top gate has less control over the channel and the two sidewall gates govern the channel. Figure 7.17c shows the plot of I_{off} versus I_{on}. It can be seen from Fig. 7.18 that at a fixed fin height the cutoff frequency increases with the increase in fin width and may reach as high as 200 GHz.

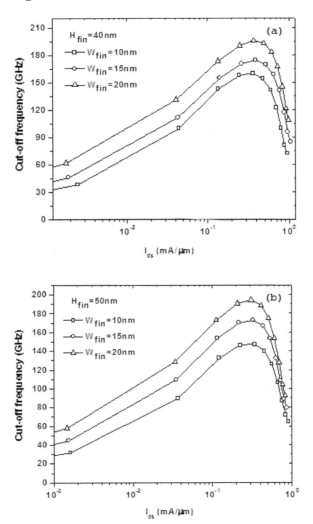

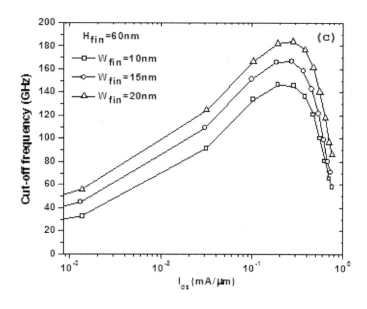

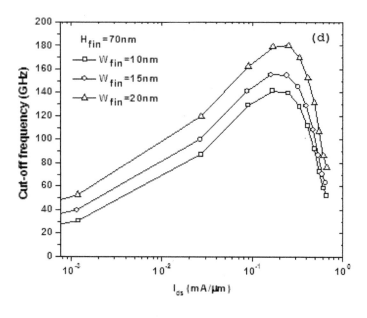

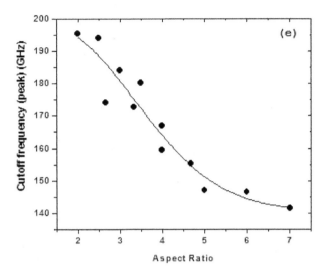

Figure 7.18 Cutoff frequency versus gate current characteristics for various fin widths at fixed fin heights of (a) H_{fin} = 40 nm, (b) H_{fin} = 50 nm, (c) H_{fin} = 60 nm, and (d) H_{fin} = 70 nm, and (e) peak cutoff frequency versus aspect ratio plot [5].

7.4 Summary

FinFETs are widely regarded as the most promising option for continued scaling of silicon-based transistors after 2010. In this chapter, we examined key process development that enable reduction of both device area and fin width necessary for technology nodes beyond 16 nm node. Because aggressively scaled finFET structures suffer significantly degraded device performance, detailed analysis of stress-induced mobility enhancement, including high stress levels, has been presented.

References

1. Smith, C. E. (2008). *Advanced Technology for Source Drain Resistance in Nanoscale FinFETs*, PhD thesis, University of North Texas.

2. Han, L. (2015). *Investigation of Gate Dielectric Materials and Dielectric/ Silicon Interfaces for Metal Oxide Semiconductor Devices*, PhD thesis, University of Kentucky.

3. Bhoj, N. (2013). *Device-Circuit Co-Design Approaches for Multi-Gate FET Technologies*, PhD thesis, Princeton University.

4. Demircioglu, H. (2013). *Modeling Layout Dependent Stress Effects for CMOS*, PhD thesis, North Carolina State University.

5. Sengupta, M. Private communication.

Chapter 8

Advanced Devices

As the conventional planar bulk silicon MOSFET reached its limits, from the sub–32 nm technology nodes, new architectures such as fully depleted (FD) are being considered. The FD silicon-on-insulator (SOI) architecture is based on adding a buried dielectric to improve the control of parasitic effects and limit MOSFET leakage currents, while remaining with the classical bulk integration. For further scaling, the ultrathin FD SOI MOSFETs have attracted special attention. Compared to PD SOI, they have additional key advantages:

- Reduction of short-channel effects (SCEs): Benefiting from the ultrathin body, the leakage paths between source and drain triggered by SCEs are suppressed, leading to limited the threshold voltage roll-off and finally to the reduction of off-state current and power. On the other hand, drain-induced barrier lowering (DIBL) can also be reduced with the film thickness shrinking. A thinner BOX also leads to smaller DIBL due to the reduction of the fringing field through the BOX and the substrate. In addition, the ideal subthreshold swing ($\sim$60 mV/dec) is achieved in ultrathin FD MOSFETs.

- Multiple threshold voltage: Another attractive feature for FD SOI devices is the back gate, which enables one to adjust the threshold voltage for low-power management. Compared to bulk silicon technology, where the threshold voltage can only

Introducing Technology Computer-Aided Design (TCAD): Fundamentals, Simulations, and Applications
C. K. Maiti
Copyright © 2017 Pan Stanford Publishing Pte. Ltd.
ISBN 978-981-4745-51-2 (Hardcover), 978-1-315-36450-6 (eBook)
www.panstanford.com

be tuned by processes such as channel implanting and gate work function engineering, tuning the threshold voltage by the back gate in FD technology is much simpler and more flexible. Wise back-gate bias also helps improving the carrier mobility and SCEs.

- Undoped channel: An undoped channel, typical for ultrathin FD MOSFETs, avoids the mobility degradation from channel doping and reduces the variability of the threshold voltage induced by dopant fluctuation. However, although ultrathin FD SOI technology shows unrivalled advantages in suppressing SCEs and exhibits high performance, it still faces some issues, such as increase of parasitic source/drain resistance, diffusion of source/drain dopants, the self-heating effect, the parasitic bipolar effect, and coupling effects. In this section, we focus on FD SOI MOS transistors on thin-film SOI for the 10 nm technology node.

8.1 Ultrathin-Body SOI

The planar bulk silicon metal–oxide–semiconductor field-effect transistors (MOSFETs) suffer from many parasitic effects:

- Current leakage in the substrate is too high.
- Doping of the channel induces a high variability of the threshold voltage.
- The electrostatic control of the gate on the channel is greatly degraded.

From the 28 nm technology node, therefore, the industry migrated to devices with thin silicon under the gate to obtain fully depleted (FD) channels. The main advantage of FD channels is the improvement of electrostatic control. Two main architectures for obtaining these types of channels are:

- A planar approach (FD silicon-on-insulator [FD SOI]) where a thin silicon film is obtained by inserting a buried oxide.
- A nonplanar approach (fin-shaped field-effect transistor [finFET]) in which the conduction channel is formed in a plane perpendicular to the surface of the substrate.

For advanced technologies, the use of FD channel transistors is mandatory to limit the effects of short channels and ensure good electrostatic control which increases performance. Two approaches currently in use are:

- FinFET or trigate architecture that exploits the third dimension to form channels with an aspect ratio (height/width) significantly greater than 1. These channels have a width of about 10 nm and are referred to as fins.
- Planar FD SOI architecture where the silicon film is thinned (<10 nm) on an insulator.

The SOI technology originates from the research on silicon-on-sapphire (SOS) in the 1960s. The SOI substrate comprises three layers: the active silicon film device layer, the buried oxide (BOX) and the silicon substrate. Depending on the thickness of silicon film, the SOI substrates can be divided into two groups, partially depleted (PD) and FD SOI. Compared to bulk silicon transistors, PD SOI technology has several advantages:

- The BOX simplifies the isolation of devices, and completely avoids the parasitic effects such as latch-up, charge sharing, and leakage between devices.
- Due to the natural isolation by the oxide, SOI devices are immunized from radiation effects (especially single-event effects due to charge in the channel).
- SOI circuits exhibit less parasitic capacitance, substrate noise, and energy consumption due to lower leakage and supply voltage.

An alternative integration technique to form the gate of transistors after the annealing for activating the source/drain regions, the so-called "gate last" on SOI will be presented. The purpose of integration gate-last on SOI is to adjust the threshold voltage of the MOSFET and to optimize the stress in the channel. This integration also paves the way for reducing the gate dielectric thickness. Nowadays, several architectures are possible make nanoscale transistors: finFET or planar transistors. Decreasing the size of transistors does not only aim to reduce the manufacturing cost, the increasing number of functionality, and speed is also improved from one technology node to another. Since the 2000s, the reduction in size alone has not been enough to improve the performance of transistors marking the end of the "happy scaling" era. For the 10 nm technology node, innovative architecture involving a buried SiGe layer to transfer stress to the channel beneficial for the n-MOSFET has been proposed [1].

8.2 Gate-First SOI

The main process steps (Fig. 8.1) for the integration of gate-first SOI technology are:

1. The starting substrate is with a thin BOX (25 nm) or a thick BOX (145 nm). The isolation is performed by shallow trench isolation (STI). Biaxial tension stress can be introduced by using substrates such as strained SOI (sSOI) with carrier transport in the crystallographic direction <110> for a (001) surface.

2. The gate stack is deposited and then etched using standard process lithography. The final gate stack includes:
   - High-k (1.9 nm HfSiON) on a plasma oxide (0.8 nm)
   - 6.5 nm PVD TiN as metal gate
   - 50 nm polysilicon

3. After the first spacer, the source and drain are formed by selective epitaxy. The source and drain are made of silicon but source and drain SiGe can advantageously be integrated for the p-MOSFET. The areas between the regions' source and drain and the channel (i.e., the extensions) are electrically connected by ion implantation and then annealed.

4. Then the second spacer is formed and the source and drain are doped and annealed for activation (~1050°C). Finally silicide (NiSi) is formed on the source and drain.

5. A nitride layer of 40 nm serving as a barrier layer to the etching of the contact is made. This layer, called the contact etch stop layer (CESL), can be of a different nature with an intrinsic tensile stress (1.6 GPa), neutral, or compressive (−3 GPa).

6. The CESL is then covered by a deposited oxide (interlayer dielectric).

7. Finally, contacts are formed followed by back-end processing.

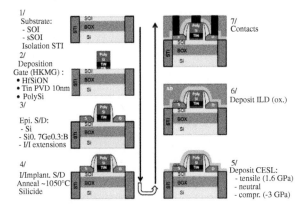

Figure 8.1 Main manufacturing steps of gate-first transistors on SOI [1].

8.3 Gate-Last SOI

The main process steps (Fig. 8.2) for the integration of gate-last SOI technology are discussed below.

The first steps are similar to the integration gate-first:

1. SOI substrate with a thin BOX (25 nm). The isolation is performed by STI.

2. High-k (HfSiON) is deposited and the polysilicon form the sacrificial gate. Unlike the integration gate-first, polysilicon is deposited here directly on the high-k and not on the metal grid.

3. After the first spacer, the source and drain are formed by selective epitaxy. Then the second spacer is formed, followed by the source and drain are doped. The activation annealing (~1050°C) is then performed. Finally silicide is formed on the source and drain.

4. A CESL of 20 nm is then deposited on the inter layer dielectric.

5. Then part of dielectric is etched from CESL: the top of the sacrificial gate is discovered by the steps of chemical-mechanical polishing (CMP).

6. The polysilicon gate is removed by tetramethyl ammonium hydroxide (TMAH) solution.

7. The gate metal (TiN PVD) as well as encapsulating metals (TiN/W) are deposited in the cavity thus formed.
8. Metal deposited outside the gates of cavities is removed by CMP.
9. Finally, the contacts are formed.

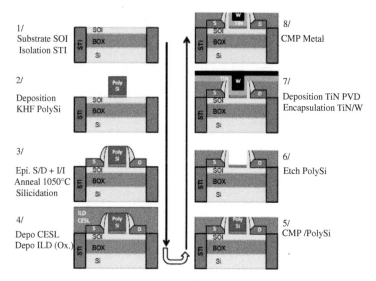

Figure 8.2 Main manufacturing steps of high-*k* gate-last transistors on SOI [1].

8.4 3D SOI n-MOSFET

In the following, a SOI n-MOSFET with a body contact is simulated using VictoryDevice.

A body contact is used to suppress the kink effect. The device structure is created using DevEdit and exported to VictoryDevice. Use of a body contact on SOI n-MOSFET suppresses the kink observed due to impact ionization in the I_d–V_d characteristics. The effect of lattice heating is also shown. Device simulation involves the following steps:

- Definition of 3D SOI MOSFET structures using DevEdit
- Study the effect of body contact in 3D MOSFET
- Calculation of lattice-heating effects
- Simulation of impact ionization effects

DevEdit defines the basic SOI n-MOSFET structure first. Although the geometry of the device is essentially 2D, it is simulated in 3D so that the results can be directly compared with those from a 3D MOSFET with a body contact. Using a tetrahedral mesh, the first MOSFET structure is saved and is shown in Fig. 8.3, which is then imported into VictoryDevice.

Figure 8.3 DevEdit-generated SOI n-MOSFET structure with no body contact.

In VictoryDevice, the statement ELECTRODE SUBSTRATE is used so that the substrate of the device is recognized as an electrode, which will be held at ground potential. The succeeding CONTACT statement assigns to the gate the work function of n-type polysilicon. The MODELS statement is used to specify the particular models to be used during the simulation. In the simulation, CVT, a general-purpose mobility model, including concentration, temperature, parallel-field, and transverse-field dependence, Shockley–Read–Hall (SRH) recombination, AUGER, Auger recombination, and bandgap narrowing (BGN) in the presence of heavy doping were used. In addition to these models, the IMPACT SELB statement specified that the Selberherr model for impact ionization to be used. Once the contacts, models, and methods have been specified, the simulation begins with a SOLVE INIT to initialize the solution at zero bias. Next a small initial bias is applied to the drain, and then the gate is brought up to its operating bias of 3 V. The bias on the drain is then ramped from 0.2 V up to 4 V– and the I_d–V_d characteristics was obtained and is shown in Fig. 8.4.

After this, the entire simulation is repeated for this structure, but this time considering lattice-heating effects. The CONTACT and MATERIAL statements assign to the gate contact the work function, heat capacity, and thermal conductivity of n-type polysilicon. The THERMCONTACT statement defines a thermal contact at the substrate, and specifies that it will be maintained at 300 K. Once the gate has been biased, lattice temperature simulation is activated by the statement MODELS LAT.TEMP. The drain is then biased as before, and the I_d–V_d characteristics is obtained as shown in Fig. 8.4. The second DevEdit run defines the MOSFET with an additional body contact besides the gate. The difference between this structure and the first one can be seen by comparing the Figs. 8.3 and 8.5.

As with the first structure, simulations are run with and without lattice-heating effects. With the addition of the body contact (at 0 V), however, the device is much more stable. The I_d–V_d characteristics are shown in Fig. 8.4 for comparison. This comparison clearly shows that the addition of the body contact suppresses the kink which is due to the impact ionization.

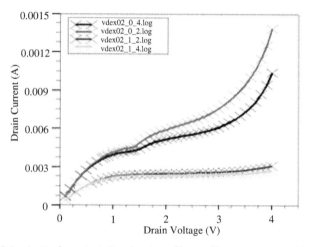

Figure 8.4 I_d–V_d characteristics showing effects of body contact and heating.

Figures 8.6 and 8.7 compare the impact ionization patterns for the two device structures. Figures 8.8 and 8.9 compare the heat fluxes as well as other lattice-heating effects for the two device structures.

Figure 8.5 SOI n-MOSFET with a body contact.

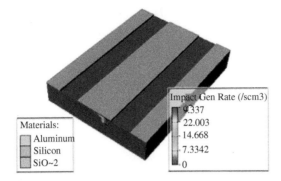

Figure 8.6 Impact ionization when body contact is absent.

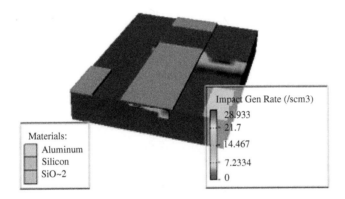

Figure 8.7 Impact ionization when body contact is present.

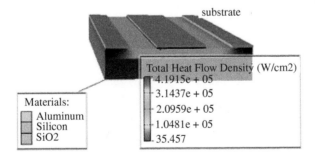

Figure 8.8 Total heat fluxes when body contact is absent.

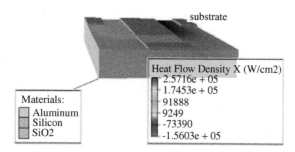

Figure 8.9 Total heat fluxes when body contact is present.

8.5 TFT

Polysilicon thin-film transistor (TFT) technology, an alternative to amorphous silicon technology has shown very high potential for high-performance, low-power, small and medium-size flat-panel mobile displays. Interest in polycrystalline silicon as the active material of TFTs has increased significantly in recent years. The polysilicon TFT technology has demonstrated its capabilities and compatibility with variety of substrates. TFT devices became the most important device for active matrix liquid crystal display (AMLCD) development. Currently, demand for TFT devices not only comes from the AMLCD industry but also from other emerging display industries such as organic light-emitting diode (OLED). Recently, TFT devices are also being considered for development of other large-area electronic systems such as bio and image sensors, printing actuators, and radio-frequency (RF) and wireless modules. The cross-sectional view of a typical self-aligned polysilicon TFT device is shown in Fig. 8.10.

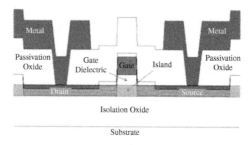

Figure 8.10 Cross-sectional view of a top-gate self-aligned polysilicon TFT [2].

Next, the design and characterization of polysilicon TFT electronics for large-area systems on flexible platform will be discussed. An integrated process and device TFT simulations performed using ATHENA and ATLAS is shown. ATHENA is used to construct the geometry and doping of a TFT device. The starting substrate is defined as silicon dioxide to emulate the flat-panel display on glass. The transistor is simulated with a metal gate at the bottom and a gate insulator made from oxide and nitride. A lightly doped silicon layer is deposited to act as the channel region. A heavily doped layer is placed on top to form the source/drain regions. Single-crystal silicon or polysilicon could also be used. Metal for the source/drain contacts is then applied. Then etch back through the metal and heavily doped silicon is done. Some of the lightly doped silicon layer is also removed. The final etch separates the source and drain. Electrodes are defined in ATHENA for use in ATLAS. The cross-sectional view of simulated TFT device is shown in Fig. 8.11.

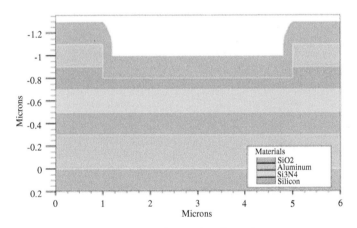

Figure 8.11 Cross-sectional view of a simulated TFT device.

In simulating TFT devices the most important requirement is setting of the defect statement specifying the density of states in the semiconductor bandgap. The defect states are specified as donor-like and acceptor-like and as tail and midgap Gaussian states. ATLAS is used to simulate the gate voltage bias from −20 V to +20 V with the drain at +10 V. The material statement is used to set the material constants of the semiconductor to those of amorphous silicon. The interface step defines an interface charge on each semiconductor/insulator interface. It is possible to vary this charge by position using the bounding box parameters on the interface statement. A constant mobility as defined by the material statement is used and SRH recombination is included. In order to simulate reverse leakage the band-to-band tunneling model is included.

In simulation, the first step in the solve sequence is to ramp the drain voltage up to +10 V. Then the reverse gate voltage ramp is applied. The gate voltage is ramped up to −20 V. The drain current increases during this ramp due to tunneling current in the TFT. The final step is the forward gate voltage sweep. This is done by first loading the solution saved with zero gate voltage. Then the gate is ramped to +20 V. From the data, the extract syntax is used to get the subthreshold leakage slope. The slope is typically 0.5–1.5 V/decade which is sensitive to the defect distribution in the semiconductor. The forward/reverse gate voltage characteristics of the TFT device are shown in Fig. 8.12.

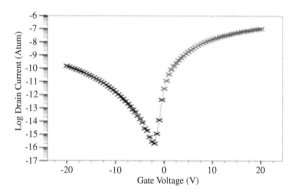

Figure 8.12 Forward/reverse gate voltage characteristics of a TFT device.

In the following example, the TFT structure is formed using ATLAS syntax, and material and model parameters for passivated a-Si are included to obtain the forward I_d–V_g characteristics. When silicon is deposited on an amorphous surface (normally SiO_2), small Si crystalline clusters or so called grains are formed over the entire surface area. Unlike single-crystal silicon, polysilicon films consist of a number of grains (crystallites) of various sizes and different orientations. The atomic disorder and unsaturated Si atom bonds (dangling bonds) at grain boundaries, and sometimes within the grains, are directly responsible for transport properties of polysilicon films. Figure 8.13 shows the device structure for simulation of grains and grain boundaries in polysilicon TFT. ATLAS and SPISCES have been used to model the influence of grain boundaries on the device. The grain and grain boundary are created by defining two regions in the polysilicon. Each region then has its own material and defect statements that define the properties within the grain and the grain boundary.

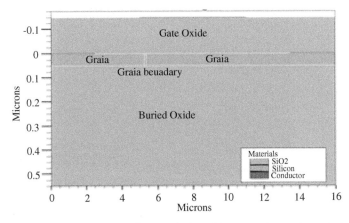

Figure 8.13 Device structure for simulation of grains and grain boundaries in a polysilicon TFT.

In TFT simulation, the defect statement is most important. It is used to define a continuous density of trap states in the silicon and the relevant trapping cross sections. In ATLAS, it is possible to specify an energy-dependent distribution of traps in two ways and comparison of discrete and continuous distribution of traps can be made. The DEFECTS statement allows the user to specify a fixed density of states versus energy distribution. This would normally comprise one Gaussian and/or one exponential energy distribution of traps across the bandgap. The capture cross sections are taken from the parameters in the DEFECTS statement. The acceptor capture cross section is defined by the Gaussian acceptor value and the donor capture cross section is defined by the Gaussian donor value. ATLAS allows the user to use the C-interpreter to specify the density of states versus energy distribution in which the acceptor and donor densities are defined by two exponential equations each of which begin from the conduction and the valence bands. The effects of a discrete and continuous distribution of traps on I_d-V_g characteristics are shown in Figs. 8.14 and 8.15.

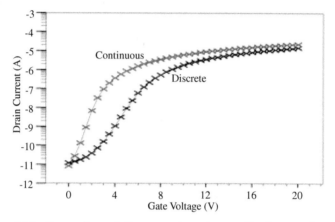

Figure 8.14 Comparison of discrete and continuous distribution of traps on I_d-V_g characteristics.

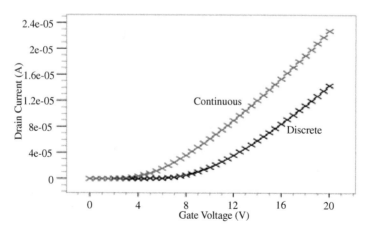

Figure 8.15 Comparison of discrete and continuous distribution of traps on I_d-V_g characteristics.

8.6 HEMTs

Current wireless communication systems face several technology challenges, despite the steady development in the last decade. One of the key technologies that made this evolution possible was the gallium nitride–based high-electron-mobility transistor (HEMT). It offers a high current density at RFs that, combined with its high breakdown voltage, makes it an excellent choice for high-power amplifiers. It is finding applications in RF communication systems, automotive electronics, and sensors. Due to higher low-field electron mobility, GaAs devices are faster and better compared with conventional silicon devices. Due to lower saturation field, GaAs can be semi-insulating with a high-energy gap. GaAs shows less parasitic capacitance resulting in further speed advantages than silicon. As GaAs operates at higher voltages, it is more also useful for RF power amplifiers. However, migration from silicon to GaAs is difficult for several reasons. In Si technology, silicon dioxide, SiO_2, is used for masking, which is not possible in GaAs technology.

The first GaN-based transistors were realized in the early to mid-1990s. GaN HEMT and other wide-bandgap semiconductor devices are being investigated for applications in communications. Using GaN technology over currently fielded gallium arsenide (GaAs) could result in a tenfold increase in power density at identical frequencies. GaN devices offer superior material properties for high-power, high-bandwidth applications, especially at high voltages where current GaAs technology is unsuitable. Diamond has the highest thermal conductivity of any known substance and can be easily grown on large wafers.

Understanding the thermal and electrical properties will allow for optimization of the GaN transistor structure, the geometry of the diamond substrate and prediction of thermal conductivity across layer interfaces. If a high thermal conductor such as diamond can be utilized in a GaN HEMT to reduce channel temperatures, both improved device performance and most importantly component lifetimes may be increased by several orders of magnitude. Popular substrate materials currently used for GaN HEMTs include sapphire,

Silicon carbide (SiC), silicon, and aluminum nitride (AlN): Each substrate choice has been proven with individual successes and challenges. There is a requirement for simulation tools to accurately predict device performance prior to fabrication because of the high inherent cost involved.

An HEMT is a FET that operates very similar to a metal–semiconductor field-effect transistor (MESFET). Electron flow across the carrier channel from source to drain is modulated by changing gate voltage. The main difference between a MESFET and a HEMT is the device structure. HEMTs use different compounds grown in layers to optimize and extend the performance of the MESFET. The different layers form a heterojunction. Figure 8.16 shows the basic HEMT structure.

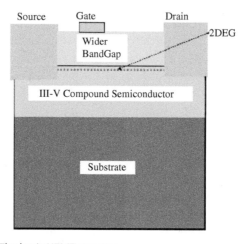

Figure 8.16 The basic HEMT structure.

The device under consideration is an AlGaN/GaN HEMT. The 2D structure is obtained by ATHENA and the stress distribution is simulated by VictoryStress. The simulation produces stress distributions via the stress liner made of nitride (tensile or compressive). Note that the stressor in this case study is used to illustrate strain polarization, on top of lattice and spontaneous polarization. The polarization model that supports dependency on a loaded strain tensor has been used. The model is enabled by the TEN.PIEZO flag in the MODELS statement.

The simulation steps are:

- Construction of the heterojunction structure using the ATHENA syntax

- Stress simulation, material and model parameter specification
- Simulation of I_d-V_g characteristics

The main idea of GaN-based power devices is to use epitaxial strain to create a 2DEG. The polarization model that supports epitaxial strain due to lattice mismatch is enabled by the CALC. STRAIN flag on the MODELS statement. When both models are set in the simulation both the imported strain and the lattice mismatch calculation are accounted for. The lattice and imported strain-dependent components of polarization can be independently scaled using the scale factors TENSO.SCALE and PIEZO.SCALE on the MODELS statement. When enabling the TEN.PIEZO flag spontaneous polarization is automatically included in the calculation. The size of the spontaneous component can be scaled using the PSP. SCALE parameter of the MODELS statement. The device under consideration is a 3D AlGaN/GaN HEMT shown in Fig. 8.17. Figure 8.18 illustrates strain polarization on top of lattice and spontaneous polarization. The stress distribution is simulated by VictoryStress and stress distributions via the stress liner made of nitride as shown in Fig. 8.19.

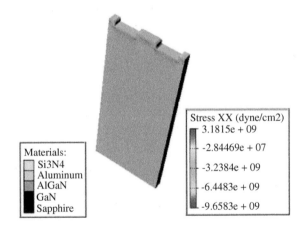

Figure 8.17 A 3D AlGaN/GaN HEMT structure.

After the initial solution is obtained the drain voltage is ramped to 1 V, then the I_d-V_g characteristic is extracted from $V_g = -5$ to $V_g = 1.0$ V. The effect of strain is seen on I_d-V_g characteristic. For illustration purpose we combined, in this case study, different intrinsic stresses (compressive or tensile), leading to different stress distributions and signs in the channel and the source and drain regions. As a result I_d–V_d characteristics change accordingly, as shown in Fig. 8.20.

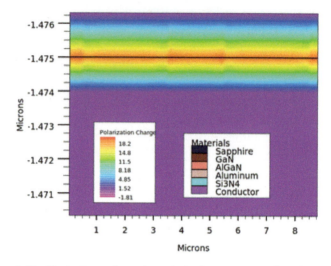

Figure 8.18 Illustration of strain polarization on top of a lattice and spontaneous polarization in a GaN HEMT.

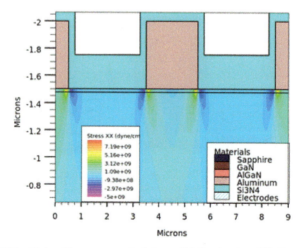

Figure 8.19 VictoryStress-generated stress distribution in a HEMT.

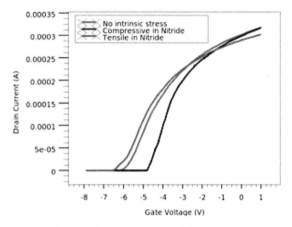

Figure 8.20 I_d–V_g characteristics of a stressed GaN HEMT.

8.6.1 Thermal Optimization Using a Flip-Chip Structure

The thermal conductivity of GaN is a challenge to overcome. While able to support high temperature operation, GaN by itself is unable to sufficiently remove the heat generated during intended device operation. Increased thermal resistance and shortened device component lifetimes are just two of the many undesired effects of high device temperatures. Removing the heat generated during operation could dramatically increase device performance, reliability and lifetime. Substrate selection for GaN HEMTs has primarily focused on sapphire and SiC due to their availability and ease of growing GaN onto these substrates.

In this case study, 2D electrothermal simulations with heat sink structures are under taken. Although the superiority of GaN HEMT device characteristics has been demonstrated, the self-heating effect has hindered the production of high-power and high-speed GaN-based switching devices. This effect can be significantly reduced by the cost-effective heat sink approach. In this case, in order to understand and control the self-heating effect, a GaN HEMT with a flip-chip concept is simulated, and device characteristics are compared versus a normal structure.

For HFETs, the GaN/AlGaN epitaxial layers have been grown on either sapphire or SiC substrates. Although sapphire has the advantage of lower cost and availability in larger wafer sizes, its poor thermal conductivity (0.3 W/cm-K) limits the achievable powers due to severe self-heating. The self-heating effect can be significantly reduced by flip-chip mounting the devices onto highly conducting substrates such as AlN (1.8 W/cm-K).

The typical GaN HFET flip-chip structure in this example is an $Al_{0.25}Ga_{0.75}$ N-GaN HFET on a sapphire substrate. The Silvaco software, a physics-based modeling program, was utilized to model and simulate the GaN HEMT on a sapphire. The structure consists of an AlN layer as a heat sink, a 2.7 nm undoped AlGaN layer, and two GaN layers which includes 20 nm doped 1×10^{15} per cc GaN and 1 μm undoped GaN as shown in Fig. 8.21. The spaces of the source-gate and drain-gate are 2.0 μm and 2.0 μm respectively, and gate length is 1.0 μm. The spontaneous polarization of the interface of AlGaN/GaN is taken into account in ATLAS. The 2DEG was calibrated to 9×10^{12} cm^{-2} using the polarization scaling factor. The drift-diffusion transport model is used for this simulation with the modified Caughey–Thomas mobility for the low field, and the high-field-dependent mobility model is based on fitted Monte Carlo data for bulk nitride.

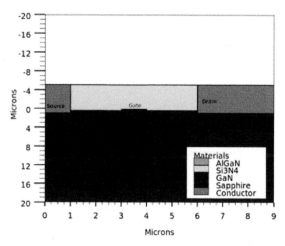

Figure 8.21 Schematic cross section of a simulated GaN HFET with a flip-chip structure.

The self-heating is a local increase of crystal temperature due to dissipated Joule electric power, this effect can significantly reduce the electron mobility and degrade device performance. DC, transient, and AC simulations are performed and compared with and without the flip-chip structure. Figure 8.22 shows the current–voltage (I_d–V_g) comparison. In Fig. 8.22, simulations for the normal structure show significant degradation of transconductance compared to the case of the flip-chip structure. Figure 8.23 shows the I_d–V_d characteristics versus gate bias. We can clearly see significant degradation of output characteristics with pronounced negative differential output conductance (NDC) region and with relatively high temperature at the drain-side gate edge due to the self-heating effect (Fig. 8.24). Degradation of output characteristics in the case of the flip-chip is significantly improved. Figure 8.24 shows the lattice temperature distribution. The NDC depends on the thermal dissipation. The normal structure exhibits stronger NDC compared to the flip-chip structure, because the sapphire thermal conductivity is smaller than AlN.

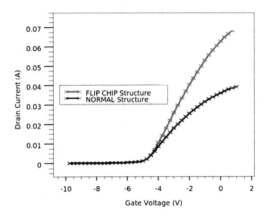

Figure 8.22 I_d–V_g comparison.

Figure 8.25 shows the self-heating impact on transient performance. In gate lag simulation, a turn-on step voltage (≤10 ns) is applied to the gate terminal (from V_g = –6 to 0 V), maintaining a fixed drain bias of V_g = 5 V. The drain–current transient versus time is analyzed. In Fig. 8.25, simulation for the normal structure clearly shows significant current collapse compared to the flip-chip structure. This current collapse can be correlated to temperature effect due to self-heating. Figure 8.26 shows RF characteristics.

Simulations indicate ~50% enhancement in f_T and f_{max} with a flip-chip structure that is in good agreement with transconductance and output simulation data. As expected the performance of the device is better using the flip-chip architecture.

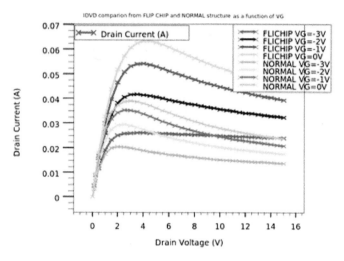

Figure 8.23 $I_d–V_d$ comparison.

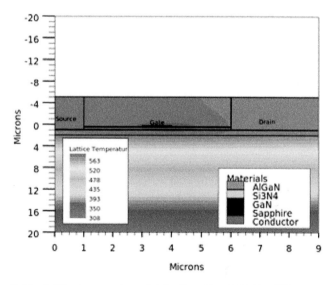

Figure 8.24 Lattice temperature distribution: V_d = 15 V; V_g = 0 V.

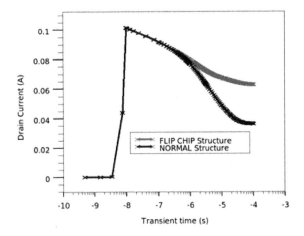

Figure 8.25 Drain current versus transient time.

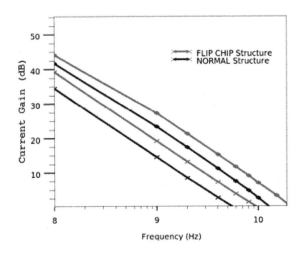

Figure 8.26 RF performance comparison.

8.7 AlGaN/GaN HFET

In the following, we take up the simulation of a normally off AlGaN/GaN HFET with p-type GaN gate and AlGaN buffer. GaN-based devices exhibit piezoelectric as well as spontaneous polarization.

Gradients of polarization charges lead to charge accumulation at heterointerfaces and a strongly induced localized 2DEG. Polarization modeling is thus critical for GaN-based devices. ATLAS provides three different polarization models for GaN and the related nitrides. The latest model TEN.POLAR calculates the piezoelectric and spontaneous polarization but also includes contributions by external mechanical strain and axial strain due to lattice mismatch. ATLAS uses specific physical models and material parameters to take into account the mole fraction of the AlGaN/GaN system.

We choose to model low field mobility using the ALBRCT model allowing the control of electrons and holes separately thus taking into account the fact that the gate is p-type. This mobility model is also a function of lattice temperature. We have selected a nitride-specific high-field-dependent mobility model. This model is based on a fit to Monte Carlo data for bulk nitride and is set by adding GANSET.N (for electrons) in the model statement. In some cases, lattice heating may be important. This typically occurs at high current operation, just like the case of power devices. The lattice heating model should be used to simulate the heat flow in the device and reproduce negative differential resistance. To enable heat flow simulation, the LAT.TEMP parameter is set on the MODEL statement. The structure was created using the ATLAS syntax. Mesh was optimized in order to get an accurate and quick simulation, paying special attention at the interfaces where induced charge from polarization are present. The structure as well as the polarization charges automatically calculated from the polarization model are shown in Figs. 8.27 and 8.28. While a Schottky-type metal on the AlGaN barrier acts as gate for normally on HEMTs, a p-type doped semiconductor as gate is able to deplete the transistor channel when unbiased, thus yielding a normally off device. The p-GaN gate transistors presented here combine the high-mobility 2DEG transistor channel with secure normally off operation, as is required for applications in power electronics. However, the required $V_t > +1$ V is often achieved by a low Al concentration in the AlGaN barrier, giving a reduced electron density in the 2DEG of the transistor channel and compromising R_{ON}. A low-A-concentration AlGaN buffer beneath the GaN channel is introduced to gain both a high electron concentration in the 2DEG and a high V.

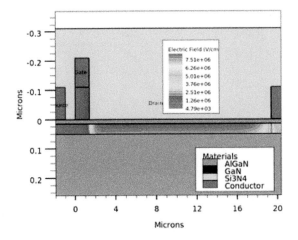

Figure 8.27 GaN FET structure used in simulation.

The simulation results of the I_d–V_g and I_g–V_g characteristics are shown in Fig. 8.29. The threshold voltage is around 1.25 V. The subthreshold leakage current drops significantly immediately below the threshold voltage; however, the drop slows down to around 4 µA/mm at $V_g = 0$ V. The leakage current is determined by the traps. The gate current in the on state (defined as $V_g = 5$ V) is around 3 µA/mm and thus around 5 orders of magnitude below the drain current.

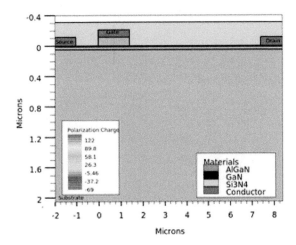

Figure 8.28 GaN FET structure showing polarization charges.

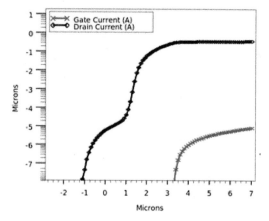

Figure 8.29 I_d–V_g and I_g–V_g characteristics.

The output characteristic shown in Fig. 8.30 exhibits a negative differential resistance due to lattice heating and simulated by solving the lattice heating equation set by the LAT.TEMP parameter of the MODELS statement. A breakdown voltage of 870 V for an 18 μm gate–drain spacing is shown in Fig. 8.31. GaN power devices are expected to prevail in high end applications over more traditional semiconductors such as Si or gallium arsenide with GaN offering up to five times the power density than that of GaAs.

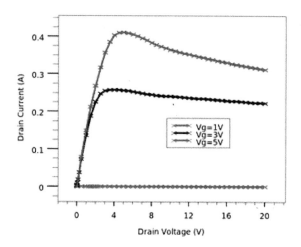

Figure 8.30 Output characteristics.

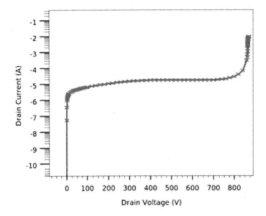

Figure 8.31 Breakdown voltage characteristics.

8.8 3D SiC Process and Device Simulation

Power MOSFETs are widely used electronic devices capable of delivering high power levels. Silicon carbide is a promising wide-bandgap semiconductor material for high-temperature, high-power, and high-frequency device applications due to its high thermal conductivity and high critical field for breakdown. Interest in silicon carbide power electronic devices has been increasing steadily with the advancement of obtaining devices with low defect concentrations. It has been shown that SiC devices have a higher power density than silicon devices. In general, wide-bandgap semiconductors offer a lower intrinsic carrier concentration, a higher electric breakdown field, a higher thermal conductivity, and a faster saturated electron drift velocity. SiC is an attractive material for high-temperature operating (>650°C) gas sensors as well as solid-state transducers such as pressure sensors and accelerometers for automotive and space industry applications using microelectromechanical systems. Although with favorable properties of SiC, the full performance of SiC devices is limited by the material quality itself and the fabrication of high temperature stable Schottky contacts and low-resistivity Ohmic contacts. Generally, contacts between metals and semiconductors play a major role in all classes of devices.

Out of various polytypes with different crystal structures with the same stoichiometry, only the 6H- and 4H-SiC polytypes are available commercially as both bulk wafers and custom epitaxial layers. SiC has equal parts silicon and carbon, both of which are group IV elements. The distance between neighboring silicon (a) or carbon atom is approximately 3.08 Å for all polytypes. Both 6H- and 4H-SiC polytypes have a hexagonal crystal structure and a bandgap in the neighborhood of 3 eV. The carbon atom is situated at the center of mass of the tetragonal structure outlined by the four neighboring Si atoms.

In the following, we demonstrate 3D trench SiC IGBT simulation using VictoryCell and VictoryDevice. The device has a low doping long drift region of about 160 µm which will lead to high breakdown voltage. To create the 3D structure we used VictoryCell 3D process simulator. After the process simulation is done a 3D structure is saved using a tetrahedron mesh to ensure that any shape created during 3D process simulation is well conserved and transferred to VictoryDevice for device simulation. SiC exhibits hexagonal crystal structures. As a consequence anisotropy features like impact ionization were taken into account. A 3D SiC MOSFET simulation was performed using DEVICED3D.

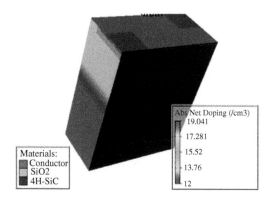

Figure 8.32   SiC transistor structure used in simulation.

8.8.1   Device Simulation

One of the features of the SiC/oxide interface is the very high density of interface states under the gate, which dominate the determination of the threshold voltage. Typical interface state density as high as $1 \times 10^{14}/cm^2/eV$ at the band edges, dropping to a near constant value of $2 \times 10^{11}/cm^2/eV$ throughout the rest of the bandgap that is more than 0.5 eV away from each of the band edges. These interface states are modeled using the **intdefects** statement. The high density of interface states effectively turn off the channel, thus creating a normally off device with a high threshold voltage.

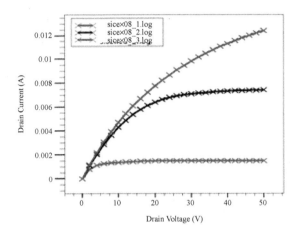

Figure 8.33   Output characteristics' simulation of SiC transistors.

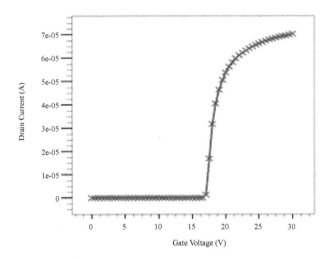

Figure 8.34   Breakdown voltage simulation of low-voltage SiC transistors.

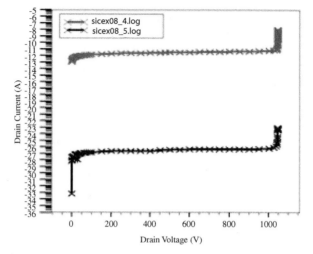

Figure 8.35   Breakdown voltage simulation of high-voltage SiC transistors.

8.9   Summary

GaN is a very promising material for high-power switching applications. However, the development cost is a serious concern for GaN manufacturers. A systematic simulation procedure has been followed for optimizing, using 2D simulation, the device geometry to obtain the maximum with minimum degradation in on resistance. Understanding the self-heating and the temperature effect in GaN HFET is an important issue because this device is a promising candidate for ultrahigh-power microwave systems, power electronics and high temperature applications. Two-dimensional electrothermal simulations for the GaN HFET with heat sink structures were also performed by ATLAS. The effects of field plate geometry and material variables in a HEMT were studied.

A normally off GaN transistor for power applications with a low on-state resistance and high breakdown strength was simulated. The combination of a p-type GaN gate with an AlGaN back-barrier yield in a sufficiently high threshold voltage for power electronic applications. At last, 3D trench SiC IGBT simulation using VictoryCell and VictoryDevice has been performed.

References

1. Morvan, S. (2013). *Transistors MOS sur films minces de Silicium-sur-Isolant (SOI) complètement désertés pour le noeud technologique 10 nm*, PhD thesis, Universite de Grenoble.

2. Jamshidi-Roudbari, A. (2010). *Design and Characterization of Polysilicon TFT Devices, Circuits and Systems for Large Area Flexible Electronic Applications*, PhD thesis, Lehigh University.

Chapter 9

Memory Devices

The semiconductor memory plays a significant role in the integrated circuit (IC) design and has in fact been one of the greatest driving forces in the advancement of semiconductor industry. The International Technology Roadmap for Semiconductors forecasts a fourfold increase in the number of bits every 4–5 years. Development in technology and design of existing memories has led the chip capacity almost doubled every two years. But features in multigigabit memory technologies do not scale as easily. In DRAM, for example, the storage capacitance must be kept constant for soft error reliability, sensing signal margin, and data retention considerations, while the transistor's threshold voltage is kept nearly the same to minimize charge leakage from the storage capacitor. Likewise, a typical Flash memory cell looks similar to a MOSFET, except that it has a dual-gate structure. Flash memories have a fixed tunnel oxide thickness for achieving 10 year retention time. High field/current stress caused by Fowler–Nordheim tunneling (during program/erase cycling) leads to tunnel oxide degradation, which eventually limits the endurance characteristics.

As a CMOS approaches the 22 nm node and below, to address the scaling challenges, new materials and approaches are required toward memory scaling with devices which use conventional CMOS materials. Lower operating voltages and faster switching can be achieved by using bandgap-engineered gate stacks, multiple

Introducing Technology Computer-Aided Design (TCAD): Fundamentals, Simulations, and Applications
C. K. Maiti
Copyright © 2017 Pan Stanford Publishing Pte. Ltd.
ISBN 978-981-4745-51-2 (Hardcover), 978-1-315-36450-6 (eBook)
www.panstanford.com

metal floating gates, thinner oxides, and tunneling as the main programming mechanisms. Floating-gate devices are commonly used as storage elements but they may find applications as logic circuit elements within CMOS logic. To circumvent the limitations of conventional scaling, the semiconductor industry incorporated strained-silicon technology to boost the performance of devices. Since strain alters several semiconductor properties, its effect on all device parameters needs to be investigated. From measurements it was observed that DRAM retention degenerates with mechanical stress, while in nonvolatile memory (NVM), retention is improved with tensile stress. CMOS memory can be divided into two main categories, volatile memory (random access memory, or RAM) and NVM, or read-only memory (ROM). Figure 9.1 shows the various types of semiconductor memory devices.

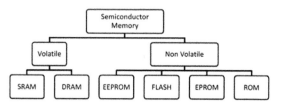

Figure 9.1 Classification of semiconductor memory [1].

In this chapter, we discuss the key challenges in designing Flash memories. Scaling the tunnel oxide thickness in nonvolatile memories (NVMs) is examined. Semiconductor memory can be divided into two categories, volatile memory and NVM. In volatile memory, the information is saved only as long as the system power is on. SRAM and the DRAM fall into this category. A typical SRAM cell is composed of six transistors and is commonly used in the cache memory of a central processing unit due to its very high switching speed between two states. It is capable of writing and reading a bit in just a few nanoseconds. However, its drawback is its big size, which results in a low density such that it is not feasible for a high-density memory array. Unlike the 6T-SRAM cell, the DRAM cell typically consists of only one transistor and one capacitor (1T–1C). The DRAM is mainly used in the main memory due to its small cell size. It can be built with a high density, which enables overall high-speed data access to recently used data. The drawback of DRAM is that the capacitor for each cell is leaky, and the cell needs to be refreshed periodically to avoid loss of the stored data. Scaling issues are related directly to the need to store a critical amount of charge on the capacitor over time.

NVM has drawn much attention over the past years due to its applications in the consumer electronics market, where memory devices with a retention time of ~10 years are desired. Flash memory is the dominant technology and offers a very high density on-chip and very low idle power consumption and low cost. It is available in a NOR and NAND architecture and it is the dominant NVM technology. In NVM, the information is stored even if the power is off. There are many technologies and technology variations that in part aim to replace the DRAM in the main memory. However, the switching speed between two states is still low compared to the volatile memory and is therefore typically used as the secondary memory storage or long–term persistent storage where a slow speed may be tolerated.

The most common NVM is Flash memory, which is based on MOSFETs with programmable floating gates embedded in the dielectrics. These so-called floating-gate devices have enjoyed enormous success as the basis for a scalable NVM technology over the past years. Flash memory has attractive features such as high density, very low idle power consumption. The biggest challenge in scaling Flash technologies is to reduce the programming voltage.

There are other NVMs that have high potential and aim for high density, high switching speed, high endurance, low power, and long data retention simultaneously. They may find applications in next-generation memory technology and those that are already matured in mass production are ferroelectric random access memory (FeRAM), magnetic random access memory (MRAM), and phase change memory (PCM). Other types of NVMs which are in early stages of investigation include resistive random access memory (RRAM), carbon nanotube (CNT) memory, molecular memory, and polymer memory.

FeRAM

Ferroelectric material is typically a perovskite material and can be polarized by an electric field where the dipoles align themselves in the same direction of the field. FeRAM is based on a metal–

ferroelectric–insulator–semiconductor (MFIS) structure and stores the data in a ferroelectric film. A change in field direction results in a displacement of the dipoles in the crystal structure of the ferroelectric material. The distribution of the charge is then also shifted. Therefore, the material shows hysteresis characteristics, which enables two stable states of the device. However, FeRAM devices face several integration challenges as compatibility with a CMOS is poor. Also the cost per bit and cell size are relatively large.

MRAM and SPRAM

The MRAM technology stores data by magnetic storage elements and uses a magnetic tunnel junction. There are two types of MRAM, (i) the conventional MRAM, where the magnetic field is generated due to the current flow in the word line and the bit line, and (ii) the advanced spin transfer torque RAM (SPRAM), where the write operation is performed by the current flow through the magnetic tunnel junction itself and a write word line is not required. In MRAM the data is stored in a magnetic state of bits and is then read by sensing the resistance. Although MRAM offers high speed, excellent endurance, and low voltage, its incompatibility with CMOS processes results in a high cost per bit.

PCM

PCM, also known as PRAM, is another potential candidate for NVM and has a 3D device structure. PCM uses a thin film typically composed of a chalcogenide material that is capable of reversible phase transitions between amorphous and crystalline phases on the application of heat. It is placed between the top electrode (bit line) and a heating element that extends from the bottom electrode (word line). The memory cell is programmed through a relatively high current that heats up the chalcogenide material and leads to a thermally induced phase change. The material changes from a low resistance to much higher resistance, depending on the state of the phase. Although PCM offers high endurance, high performance, and high density, along with promising scaling characteristics, it requires a high programming current and it is sensitive to temperature variation.

RRAM

Another important NVM is RRAM. It has drawn increasing attention due to an easier fabrication process. It is scalable to at least the 16 nm node and has potential for 3D stacking for ultrahigh density. Due to its structure, an ultrathin dielectric sandwiched between the two electrodes, its operation depends on the current filament formed or broken by application of appropriate bias. If a current filament is formed, a conductive path created between the two electrodes leads to a low-resistance state. If the filament is broken it acts as a high resistance. However, control of the formation and rupture of the filament is critical. The switching mode between low- and high-resistance states can be either unipolar or bipolar. A bipolar switching speed of 20 ns to set and 60 ns to reset with a data retention time of >10 e5 s has been reported. Other NVMs under active research need to be mentioned are CNT memory, molecular memory, and polymer memory.

Flash memory is the predominant NVM technology with the fastest-growing rate due in part to its compatibility with CMOS processes. Flash memory is based on the floating-gate device that has a structure where a gate is embedded in the oxide of a transistor in order to store charge, and thus data. Conventional devices employ typically a continuous floating gate but a discontinuous floating gate may be introduced where isolated nanocrystals are inserted in the oxide. These so-called nanocrystal floating-gate (NCFG) devices are superior in many aspects to their counterpart. NCFG devices may further be modified to use a high-k dielectric or a dual-layer structure to improve charge transfer and/or retention characteristics.

The schematic cross section of a continuous floating gate embedded in the oxide of a transistor is shown in Fig. 9.2a. The upper electrode is the control gate, and the lower electrode is the conductive floating gate. The lower oxide that isolates the floating gate from the p-type silicon substrate is called the tunnel oxide. The upper oxide, which is inserted between the floating gate and the control gate, is called the control gate oxide, or sometimes the blocking oxide. The floating gate acts as the storage node in this NVM device and is configured by applying the appropriate bias on the control gate, drain, and source for an appropriate duration. By applying a high voltage on the control gate, electrons tunnel from the channel through the tunnel oxide to the floating gate, thus programming the device. The resulting accumulated negative charge on the floating gate acts as a shield between the control gate and the channel and results in a positive threshold voltage shift of the device', as shown in Fig. 9.2b.

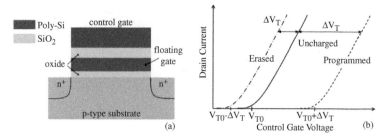

(a)   (b)

Figure 9.2 (a) Schematic cross section of a continuous-floating-gate device and (b) threshold voltage of an uncharged, programmed, and erased device. After Ref. [2].

An applied bias on the control gate in the operation mode, which is less than the new threshold voltage of the device, keeps the programmed device in the cutoff region, which leads to a minimal subthreshold leakage current. To remove the charge on the floating gate, a high negative voltage needs to be applied on the control gate such that the electrons tunnel back to the substrate. Then, the threshold voltage of the device shifts back toward the initial value V_{t0}, thus erasing the device. The amount of charge that is transferred between the floating gate and the substrate depends on several factors such as the control gate voltage, programming/erasing time, and tunnel and control gate oxide thicknesses. The materials that are chosen for the control gate and the floating gate, as well as the oxides, also contribute significantly to the charge transfer. One of the main difficulties of the continuous-floating-gate device is the formation of defects in the tunnel oxide. Defects occur after many programs and erase cycles (typically 10^5–10^6 cycles) where electrons get trapped in the oxide, thus establishing a conductive path between the floating gate and the channel. A high-quality and sufficiently thick layer (6–7 nm) requiring a high program/erase voltage across the tunnel oxide is essential for a reliable device with good endurance. Thus scaling of the continuous-floating-gate device is limited due to stress-induced leakage current (SILC) beyond the 45 nm technology node.

9.1   Nanocrystal Floating-Gate Device

Recently, nanocrystal floating-gate (NCFG) devices have drawn much attention as nonvolatile memory (NVM). A schematic cross section of this device is shown in Fig. 9.3. A discontinuous floating gate with a Si nanocrystal embedded in the oxide has been demonstrated. The nanocrystal has a diameter of just a few nanometers and an areal density of typically 10×10^{11}–10×10^{13} cm^{-2}. The tunnel oxide is also thinner (less than ~4 nm) than in the continuous-floating-gate device, resulting in a more efficient charge transfer and better scalability. The program and erase characteristics of the NCFG are very similar to the continuous-floating-gate device. It applies only for ultrathin oxides with a thickness of less than ~4 nm.

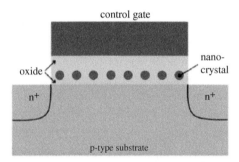

Figure 9.3   Schematic cross section of an NCFG device [2].

One of the advantages of the NCFG device compared to the continuous-floating-gate device of equivalent dimensions is the improved susceptibility to stress-induced leakage current (SILC) by storing the charge on isolated colloids of small dimensions. Figure 9.4a demonstrates how defects in the tunnel oxide lead to charge leakage of the entire floating gate to the substrate for a continuous-floating-gate device. In this case, the device changes from a programmed state to an erased state, which results in a bit error. Since these defects are very difficult to remove the lifetime of the device is practically over. In the case of an NCFG device, defects in the tunnel oxide would result in charge leakage of only the nearby nanocrystal as illustrated in Fig. 9.4b. The other nanocrystals are not affected by this and maintain their charge. Therefore, even if part of the total charge on the discontinuous floating gate is lost due to

SILC, the remaining charge may be sufficient to retain the device in a programmed state. This improves the reliability of the NCFG device compared to its counterpart and allows thinning of the tunnel oxide, which results in a more efficient charge transfer, while maintaining a retention time of ~10 years.

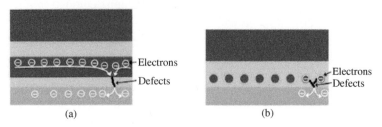

Figure 9.4 SILC susceptibility of (a) a continuous-floating-gate device and (b) an NCFG device. After Ref. [2].

The NCFG has been improved by using nanocrystal in the floating gate composed of metal instead of Si. They are superior to their semiconductor counterparts due to higher density of states, stronger coupling with the channel, better scalability and the design freedom of the work function of the metal to optimize device characteristics. High-work-function metals such as gold (Au), palladium (Pd), and platinum (Pt) with a work function of 5.10 eV, 5.12 eV, and 5.65 eV, respectively, create a deep potential well in the floating gate for efficient data retention without affecting the program characteristics of the device. However, erasing the device becomes more difficult due to the higher barrier between the floating gate and the tunnel oxide.

9.1.1 Advanced Nanocrystal Floating-Gate Devices with High-*k* Dielectrics

Use of a high-dielectric-constant (high-*k*) material as the insulator will reduce the gate leakage current by several orders of magnitude due to physically thicker (while keeping the effective oxide thickness same) layer to its counterpart, SiO_2. Recently, a wide range of structures of the NCFG device have been proposed and their schematic cross sections are shown in Fig. 9.5. The devices in Fig. 9.5b–e imply the use of HfO_2 as an insulator (HfAlO, HfSiO, ZrO_2, or Al_2O_3 are also common), while the device in Fig. 9.5f consists of a dual-layer NCFG structure. The use of a high-k material in NCFG devices such as HfO_2 offers many advantages as it is already in use in complementary metal–oxide–semiconductor (CMOS) technology, especially in the Si CMOS process with 45 nm technology and below.

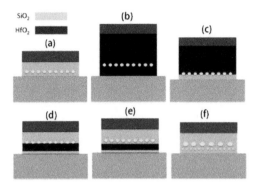

Figure 9.5 Conventional NCFG device (a) and its modified versions (b–f). After Ref. [2].

In the case of the device in Fig. 9.5b, a lower programming voltage can be achieved due to the smaller conduction band offset (i.e., barrier height) at the interface between the Si substrate and HfO_2 (bandgap 5.8 eV) tunnel oxide and the charge leakage can be reduced due to the physically larger thickness of the insulator. Also, as the effective electron mass is relatively low, an enhanced charge transfer takes place.

However, use of HfO_2 has problems associated with fabrication aspects such as, poor adhesion to Si, mobility degradation, poor thermal stability, and poor interface quality. Such fabrication issues may be solved at least partially in the device structures Fig. 9.5c–e. Device c uses HfO_2 as the control gate oxide to enable stronger coupling between the floating gate and the channel. The device in Fig. 9.5d has an additional HfO_2 layer in the tunnel oxide between the floating gate and the SiO_2 to improve both the retention time and charge transfer compared to a SiO_2-only tunnel oxide. In the device in Fig. 9.5e the tunnel oxide is composed of a symmetric $SiO_2/HfO_2/SiO_2$ stack to obtain similar benefits as the device in Fig. 9.5d with the additional feature of equal (or almost equal) program and erase characteristics. The device in Fig. 9.5f uses SiO_2 for both the tunnel and control gate oxide, but it has a dual-layer NCFG structure with the ability to improve both retention time and memory window.

Several other NVM devices have shown potential toward long retention time, high endurance and switching speed, and low power consumption.

9.2 Technology Computer-Aided Design of Memory Devices

Metal–oxide–semiconductor field-effect transistors (MOSFETs) employing programmable floating gates have drawn much attention in the semiconductor memory industry. Trap-assisted tunneling model which explains how drain voltage scaling affects channel hot-electron programming in the presence of oxide traps is discussed. Here, we consider the design of a single-layer NCFG at the device level. It is based on uniform direct tunneling and Fowler–Nordheim tunneling of electrons. Several NCFG device models have been proposed in the literature. Many use a numerical approach either by incorporating Monte Carlo simulations or by solving Schrödinger–Poisson equations, which need to be solved numerically. Parameters used in simulation are listed below:

- Gate length (nm)
- Gate width (nm)
- Control gate oxide thickness (nm)
- Dielectric constant in the control gate oxide
- Tunnel oxide thickness (nm)
- Dielectric constant in the tunnel oxide
- Control gate work function (eV)
- Floating-gate work function (eV)
- Nanocrystal diameter (nm)
- Nanocrystal density (cm^{-2})
- Bulk doping (cm^{-3})
- Ion-implanted channel doping (cm^{-3})
- Ion-implanted channel depth (nm)
- Surface interface charge (C)
- Temperature (K)

The NCFG device is created in the device simulator for the computation of the threshold voltage shift but also of the current densities, charge on the floating gate, and surface potential.

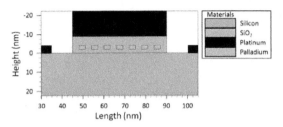

Figure 9.6 A single-layer NCFG device with 45 nm gate length designed in the device simulator. After Ref. [2].

Figure 9.6 shows a structure of a device with a gate length of 45 nm. For the control gate a high-work-function metal is preferred for better erase characteristics while program characteristics are negligibly affected. A high-work-function metal significantly reduces the probability of electron tunneling from the control gate to the floating gate, which enables vertical scaling of the control gate oxide. Platinum is chosen in the reference model which has a very high work function of 5.65 eV. Metal is also used for the nanocrystal in the floating gate because they are superior to their semiconductor counterparts. Furthermore, controlled palladium (Pd) nanocrystal deposition has been demonstrated with very good charge storage. Its relatively high work function of 5.12 eV creates a deep potential well for efficient data retention. The control gate and tunnel oxide of the NCFG device are both composed of SiO_2, which has a dielectric constant of 3.9. The p-type substrate is uniformly doped with boron at a concentration of 5×10^{16} cm^{-3}, and B ions are diffused into the channel with a Gaussian distribution at a concentration of 4.5×10^{17} cm^{-3}. The n-type source and drain are highly doped with arsenic. The geometry parameters for the device are listed in Table 9.1. Note also that a very fine mesh is chosen throughout the entire device for accurate simulation results. Especially at the material interfaces, where electrons begin to tunnel, the mesh is as thin as a hundredth of a nanometer. The mesh of the entire device consists of a total of 8790 grid points and 17,220 triangles. The tunnel oxide thickness in simulation is in the range of 2 nm to 3.2 nm. The control gate oxide thickness is 4 nm. Reducing this thickness would result in significant charge leakage from the floating gate to the control gate through the control gate oxide. The simulations are executed multiple times for different time steps and change in threshold voltage is found to be 2.0823 V for a time step of 1×10^{-7} s.

Table 9.1 Device geometry parameters used in device simulation

Parameter	Value
Tunnel oxide thickness	2–3.2 nm
Control gate oxide thickness	4 nm
Nanocrystal length	3 nm
Nanocrystal height	2 nm
Nanocrystal spacing	3 nm
Control gate length	45 nm
Control gate width	1 μm
Control gate thickness	50 nm

9.3 Process Simulation of Flash Memory Devices

A typical Flash memory process (based on conventional Si CMOS processing) and design parameters by using Silvaco ATHENA process simulator is presented below. After isolation and standard pre–gate cleaning, a 100 nm thick tunnel oxide (SiO_2) was thermally grown at 950°C with dry O_2 and diffuse time is 9 min on a p-type Si substrate in (100) orientation. Subsequently, 300 nm thick poly-Si deposited as the floating-gate charge storage layer. Approximately 100 nm thick SiO_2 at 950°C with dry O_2 and a diffusion time of 5 min, a 20 nm nitride layer, and a 10 nm oxide layer as a block oxide are deposited. A 300 nm thick poly-Si layer is deposited as a gate electrode. After gate patterning, As was implanted with dose of 1×10^{15} cm^{-2} and energies of 40 keV for shallow source/drain region contact formation, metallization, and gas anneal formation at 950°C for 50 min. The main process steps used in simulation are shown in Fig. 9.7.

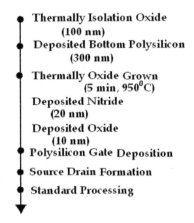

Figure 9.7 Typical process flow used in ATHENA for Flash memory process simulation.

9.4 Device Simulation of Flash Memory Devices

The structure generated by Silvaco ATHENA is characterized using the device simulator Silvaco ATLAS. The Flash memory structure is shown in Fig. 9.8. Typical ATLAS meshing commands are used to set up a MOSFET-like structure. The main difference is the presence of a layer of silicon nitride in the gate stack. The insulating layer between the nitride and the channel is termed "tunnel oxide," and the insulating layer between the nitride and the contacts is termed "blocking layer." During device simulation, three partial differential equations, viz., electrostatic potential and electron and hole concentrations were solved self-consistently using Poisson's and continuity equations. The hydrodynamic transport model was used for all simulations. During simulation, trapping and recombination of the traps were modeled using the Shockley–Read–Hall (SRH) and hot-electron injection (HEI) mechanism and the Lombardi (CVT) mobility model is used. In simulation, material properties, such as work function, bandgap, permittivity, and the electron affinity of the material were used. The ATLAS-generated SONOS structure is shown in Fig. 9.9.

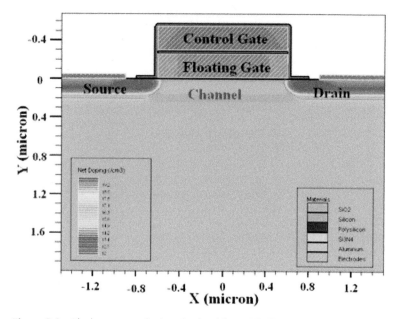

Figure 9.8 Flash memory device obtained from ATHENA process simulation.

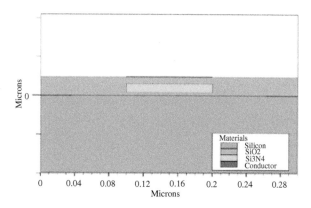

Figure 9.9 SONOS device structure used in simulation.

To study the nitride charging characteristics, the ATLAS SONOS model is invoked using the INTERFACE N.I DYNASONOS statement. The NITRIDECHARGE statement was used to set up the parameters of the nitride traps. The density of traps can be set with the parameters NT.N (acceptor-like traps) and NT.P (donor-like traps).

The parameter SIGMAT.N gives the cross section for trapping the electrons in the nitride conduction band and is set to four different values: 1.0E-14 cm², 1.0E-15 cm², 1.0E-16 cm², and 1.0E-17 cm². The parameter ELEC.DEPTH specifies the energy depth of the electron trap below the conduction band. The parameter TAU.N gives the characteristic timescale for emission from the trap level to the conduction band. The parameter SIGMAN.P controls hole capture from the valence band and is set to zero with loss of generality.

The nitride traps are charged up by ramping the bias to 16 V over 1 ns and then maintaining this for 1 ms. The stored charge as a function of time (net charge) is plotted for the four different values of SIGMAT.N. The larger the cross section, the more the charge stored. The stored charge should be the integrated difference between the current entering the nitride region from the channel and that leaving to the gate contact. As the trapping cross section increases, less charge arrives at the interface with the blocking insulator and the out current decreases and is delayed, as can be seen in the plots of the in and out current (tunneling current) at different values of the capture cross section shown in Fig. 9.10.

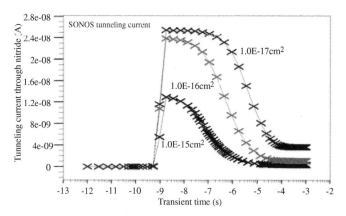

Figure 9.10 Tunneling current as a function of trapping cross section.

To view the trapped insulator charge, electron charging rate, and other relevant parameters the TonyPlot cutline tool was used. The charge stored in the nitride layer as a function of charging time is shown in Fig. 9.11. The trapped electron concentration in the insulator is shown in Fig. 9.12.

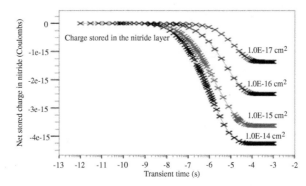

Figure 9.11 Charge stored in the nitride layer as a function of charging time.

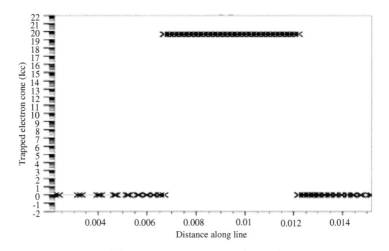

Figure 9.12 Trapped electron concentration in the insulator.

To study the threshold voltage shift, the device structure shown in Fig. 9.8 was used. However, in this structure the blocking insulator used is sapphire (Al_2O_3) and the tunneling insulator is silicon dioxide. It is necessary to make this material into a wide-gap semiconductor and set other parameters on the MATERIAL statement. The DYNASONOS model is selected on the first INTERFACE statement. Because of the relatively low conduction band offset between the silicon nitride and the sapphire, the INTERFACE S.S THERMIONIC statement is used to model any thermionically emitted current leaving the device. This current will not appear in the SONOS out current but in the gate conduction current. In the NITRIDECHARGE statement, we set up the relevant parameters and also specify PF.BARRIER. This sets the barrier height for the Poole–Frenkel detrapping model, which is enabled by the PF.NITRIDE flag on the MODELS statement.

The gate bias is ramped and the device charged for 1 s. A structure file is saved out at various preset charging times. The gate stack layers are treated as insulators in this case in order to make the simulation simpler. The charge trapped in the nitride layer affects the threshold voltage in the same way. After calculating the threshold voltage for the set of gate biases and at each of the preset charging times, the threshold data is plotted, as shown in Fig. 9.13. The threshold voltage increases with gate bias and also with charging time. For a charging time of 1.0 s there is a roughly linear relationship between gate bias and threshold voltage (0.65 V of threshold voltage change for every 1.0 V of bias voltage change).

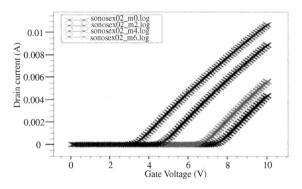

Figure 9.13 Threshold voltage (drain current) as a function of gate biases.

Here, we study how the position of the trapped charges in the silicon nitride varies with time as the device is charged up and then erased by changing the gate bias. The device has a 2 nm thick tunneling layer, a 5 nm nitride layer, and a 5 nm blocking oxide layer. The NITRIDECHARGING statement sets up a density of $10^{20}/\text{cm}^3$ of electron traps, having a relatively high capture cross section of 10^{-10} cm^2. A bias of 18 V is applied to the gate and the device is charged up. The state of the structure is saved out to a structure file at certain values of charge time. It is charged for 1.0 s, after which SOLVE INIT is called, with the SONOS parameter specified. After this, negative bias is applied to the gate, ramping for 1 ns, and then the device is erased for 1.0 s. The structure file is generated at preset intervals in order to allow the user to inspect the progress of the erasing. The integrated net stored charge in C/μm is obtained using the PROBE statement and this quantity can be plotted as a function of time. When the device is charged, the largest contribution is from the tunnel current into the nitride conduction band.

For an erase, it is the tunneling direct from the traps to the channel. This is illustrated by doing a second erase, with the only change being the parameter ELEC.DEPTH in the NITRIDECHARGING statement. This is increased from 1.5 eV to 2.0 eV in order to reduce the direct trap-to-channel tunnel current (Fig. 9.14). It can be seen from the plot that the erase takes longer with deeper traps. The TonyPlot cutline tool may be used to see the vertical profile of the trapped nitride charge through the gate stack at various time points of the charging/erasing.

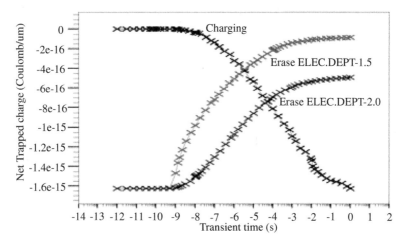

Figure 9.14 Charging and erasing cycle.

For the retention characteristics studies a SONOS device with a dielectric stack of 4 nm of silicon oxide, 5 nm of silicon nitride, and 10 nm of sapphire (Al_2O_3) is used. The device is charged for 1 ms with a gate bias of 18 V before the SOLVE INIT SONOS statement resets all electrode biases to zero. The device is simulated for up to 10 years, with progressively increasing step sizes. It saves the structure files at preset time points, so the change in charge profile and the electron charging rate as time progresses can be observed. Figure 9.15 shows the TonyPlot cutline for trapped charges through the gate stack after 3×10^8 s.

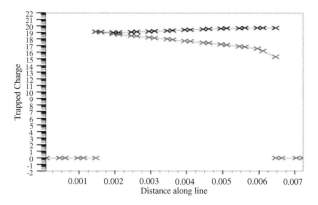

Figure 9.15   Trapped charge profile through the gate stack.

9.5   State Transition and Single-Event Upset in SRAM

Another important area of concern for nanoscale SRAM is increased susceptibility to radiation-induced soft errors. Soft errors in the form of both single-event upsets (SEUs) and SRAM array multibit fails represent a reliability concern for the memory designer. The two primary sources of soft error–inducing radiation are terrestrial radiation and radioactive isotopes within materials used in the IC fabrication process. High-energy cosmic radiation interacting with the earth's atmosphere results in a flux of neutron particles with a large range of energies extending to several 100 MeV. This process produces a charge cloud of electron–hole pairs that, when in close proximity to one or more sensitive neighboring circuit nodes, may result in a single- or multibit error.

A normal memory-state transition due to an SEU caused by an energetic particle strike in the SRAM cell is studied in the following example. In simulation, a 3D SRAM cell is created using DevEdit. Using a tetrahedral mesh, this structure is then exported to a file from which it is imported into VictoryDevice. As usual, in VictoryDevice, the first tasks are to define the models and material parameters for the simulation. The MODELS statement is used to specify the particular models to be used during the simulation. Models used in the simulation are SRH recombination with concentration-dependent carrier lifetimes; Auger recombination; mobility models, including concentration, temperature, parallel-field, and transverse-field dependence; bandgap narrowing in the presence of heavy doping; and Fermi–Dirac statistics. CONTACT statements are then used to define the work functions of the electrodes.

Once the models and contacts have been defined, the cell is brought up to its operating condition, first initializing the solution at zero bias. The cell is then brought at its operating condition. In one of its memory states biases on the emitter and anode electrodes were ramped up to 1.2 V, and ramping the bias on the drain electrode up to 1.2 V flips the memory state of the cell. The SINGLEEVENTUPSET command is used to specify the charge track of the impinging particle. VictoryDevice simulation is performed to show an SEU caused by an energetic particle strike, which strike forces an unplanned state transition in the cell.

The transient simulation is carried out in stages because the phenomena occurring in each stage occupy a different range of timescales, initially beginning with a step of 2.5×10^{-15} s and gradually increasing the step size up to a maximum of 2.0×10^{-13} s. During the peak of the SEU pulse, a constant time step of 2.0×10^{-13} s is used. The peak lasts roughly from 2.0×10^{-12} s to 7.0×10^{-12} s after the start of the transient. Following the particle strike itself, the simulation begins tracking the response of the device. A gradually increasing time step is used again, beginning with a step size of 2.0×10^{-13} s and increasing up to a maximum of 2.5×10^{-11} s. This part of the simulation continues until 3.0×10^{-10} s after the start of the transient. The contours of the potential on the surface of the device are displayed using TonyPlot3D. Figure 9.16 shows the state of the cell before the particle strike, while Fig. 9.17 represents the state afterward.

Two plots are generated using TonyPlot: One compares the various currents involved in setting and flipping the memory state, and the other shows the voltages and currents during the state flip. Once the transient simulation is complete, two more plots are generated. One compares the memory-state flip induced by the particle strike with the flip induced by raising the drain voltage. The other compares the voltages, currents, and carrier generation rate occurring during the transient.

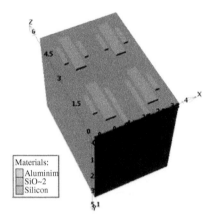

Figure 9.16   SRAM cell before the particle strike.

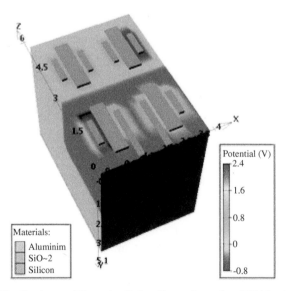

Figure 9.17   Contours of the potential on the surface of an SRAM cell after the particle strike.

9.6 Nanoscale SRAM

The widespread accessibility of multiple computing platforms available today ranging from hand-held and portable devices to mainframe supercomputers has been made possible by the reduced cost per memory bit and logic gate with each technology generation. SRAM remains the most cost-effective embedded memory solution for many such applications and could be made possible by continued advances in CMOS device scaling. However, increasing variability of the CMOS device characteristics have become the most significant problem facing future nanoscale SRAM. A number of obstacles exist to the continued use and scaling of SRAM designs beyond 32/28 nm. These include increased variation, reduced noise margins, increased standby leakage, and reliability detractors such as negative-bias temperature instability (NBTI) and radiation-induced soft errors.

A predictive technology model needs to be developed for scaled SRAM in the 90–22 nm technology nodes. The 6T SRAM cell design has so far been successfully scaled in bulk and silicon-on-insulator (SOI) technologies, the trend in 6T bit cell area is expected to continue beyond the 28/32 nm node. This continued trend in area reduction is accompanied by the well-known consequence of increased variability associated with the reduced channel area. Because of the use of narrow devices in the SRAM cell environment, the variation associated with random dopant fluctuation is a dominant variation mechanism and a major concern for future SRAM designs. Dopant fluctuations in nanoscale SRAM devices may be attributed to both random and nonrandom components. Cell layout topology, process scaling, and pushed design rules used in dense SRAM bit cell designs can influence the susceptibility to nonrandom mismatch in present and future nanoscale SRAM devices.

Here, challenges associated with future SRAM bit cell design are discussed, and the geometric variation sources which can contribute to within cell mismatch in the highly scaled array environment are examined. Because of its advantage in density, the type 4 topology remains the dominant cell design in the industry and has been successfully migrated across technology nodes (90 nm to 32/28 nm) in both bulk and SOI. As SRAM bit cell scaling continues below 32 nm, it will place increased demands on the alignment and printed dimension tolerances. Alignment-related mismatch sources is an important consideration in future SRAM cell design. The Synopsys simulation tools and simulation environments have been used to specifically examine the scaling limitations and challenges for the SRAM cell design.

Figure 9.18 illustrates four alignment driven sources that can introduce nonrandom sources of mismatch, (a) transverse or lateral straggle in SiO_2, (b) polysilicon interdiffusion-driven counterdoping, (c) lateral ion straggle from the photoresist, and (d) photoresist implant shadowing. Of these four mechanisms, (a) and (c) originate from higher energy well formation implant conditions used in bulk CMOS processes, while (b) and (d) are consistent with both bulk and SOI process technologies. These mechanisms and their impact on the SRAM devices are discussed below.

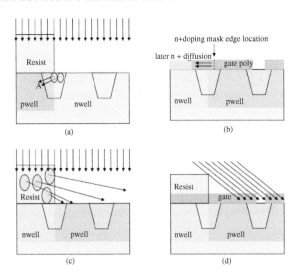

Figure 9.18 Schematic depiction of four alignment-sensitive sources of potential nonrandom mismatch in SRAM devices. (a) Lateral straggle within SiO_2, (b) lateral counterdoping in gate polysilicon, (c) lateral straggle from resist sidewall, and (d) halo shadowing. After Ref. [3].

The potential impact of transverse straggle in the SRAM cell devices arises from the aggressive n+/p+ space used in the cell to gain density. Lateral ion scattering in the shallow trench isolation oxide from the higher energy well implants can counterdope the adjacent well edge (e.g., point A in Fig. 9.18a). An NWELL mask misalignment of 30 nm is sufficient to create a substantial counterdoping path between the source and drain of the adjacent partial diffusion (PD) NMOS device, as shown in Fig. 9.19b.

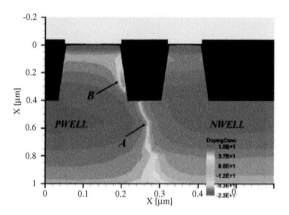

Figure 9.19 SProcess simulated well contours showing effects of transverse straggle in SiO_2 on the adjacent PWELL with 30 nm misalignment of the NWELL resist using 45 nm pushed rules. The area labeled A is a normal PWELL/NWELL boundary, and area B is a counterdoped (n-type) region in PWELL, resulting from phosphorus lateral implant straggle in STI. After Ref. [3].

Polysilicon interdiffusion is also of significant concern with scaling as n+/p+ space is aggressively pushed. The practice of using a poly predoping step is commonly used to insure the n+ polysilicon is degenerately doped. The alignment of this predoping mask as well as the n+ and p+ source drain implant masks must be carefully placed to avoid diffusion-induced counterdoping of the gate above the channel region of the complementary device, as shown in Fig. 9.20.

The physical mechanism of lateral dopant straggle stemming from the high-energy implant species in the photoresist is well known. Depending on the implantation species and acceleration energy, this mechanism can impact devices in proximity to the well edge. The amount of near-surface doping is proportional to the dose of the high-energy implant. As shown in Fig. 9.21, using implanted B11 energy of 200 keV with a dose of 3×10^{13} at/cm^2, the near-surface doping is found to be a function of the distance from the resist sidewall. As the surface concentration is a function of the distance from the resist sidewall, there is alignment sensitivity for the SRAM devices as higher channel doping levels can introduce a threshold voltage offset in the SRAM devices.

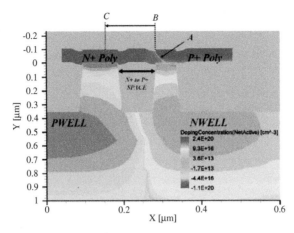

Figure 9.20 Cross-sectional simulation illustrating the concern with poly-interdiffusion across the narrow n+/p+ space in the dense SRAM environment with type 4 cell topology. Region A shows the phosphorus encroachment over the channel region of the pull-up PMOS device altering the PMOS gate work function and threshold voltage. After Ref. [3].

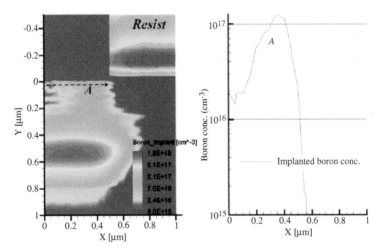

Figure 9.21 Doping contour plot following an atomistic Monte Carlo simulation of the PWELL deep implant (left). Variation in boron concentration across the silicon surface as a function of proximity to resist edge is shown (right). Doping profile taken at a depth of approximately 50 nm. The resist is located from 0.5 μm to 1 μm on the X axis. Boron lateral straggle emanating from the resist sidewall region during a deep PWELL implant results in near-surface concentration variation across the PD NMOS channel region (A). After Ref. [3].

9.7 Summary

The memory is a growing field in the IC design and intense research efforts are being made to keep this trend by improving existing technologies and finding new solutions for the future. This chapter provided an overview of the current volatile and NVM technologies. NCFG memory devices and state transition and SEUs in SRAM have been studied via technology computer-aided design simulations.

References

1. Aghoram, U. (2010). *Impact of Strain on Memory and Lateral Power MOSFETs*, PhD thesis, University of Florida.
2. Schinke, D. J. (2011). *Computing with Novel Floating Gate Devices*, PhD thesis, North Carolina State University.
3. Mann, R. W. (2010). *Interactions of Technology and Design in Nanoscale SRAM*, PhD thesis, University of Virginia.

Chapter 10

Power Devices

Smart power technologies, integrating high-voltage CMOS transistors with standard low-voltage CMOS cores, are finding an increasing use in the area of automotive applications, switching power supplies and amplifiers, and devices operating in industrial environments where the supply voltage busses are in the 12 V to 50 V range. In high-voltage applications double-diffused MOS (DMOS) transistors are more attractive than conventional MOS structures in integrated circuits, since the drift region between the drain and the active channel allows high-voltage biasing. In contrary to the DMOS, a transistor which operates in the vertical direction, the LDMOS transistor is lateral orientated and thereby length and width dependent. Currently 0.7 μm CMOS technology with the addition of 100 V n/p DMOS and 80 V NPN/PNP BJTs is in use. Process technology need to be developed to allow applications up to 100 V. However, in some applications maximum voltages of no more than 40–50 V are required, thus allowing much smaller devices than the existing 100 V devices.

10.1 LDMOS

Power metal–oxide–semiconductor field-effect transistors (MOSFETs) are solid-state switches with high power-handling

Introducing Technology Computer-Aided Design (TCAD): Fundamentals, Simulations, and Applications
C. K. Maiti
Copyright © 2017 Pan Stanford Publishing Pte. Ltd.
ISBN 978-981-4745-51-2 (Hardcover), 978-1-315-36450-6 (eBook)
www.panstanford.com

capability that evolved from complementary metal–oxide–semiconductor (CMOS) technology. The process of increasing the blocking voltage capability of a lateral MOSFET structure led to the development of a diffusion MOSFET. In an n-type lateral drain metal–oxide–semiconductor (n-LDMOS) structure the source and drain n+ regions are self-aligned with the polygate. The p-base region is then driven in deeper than the source, which is known as the double-diffusion process. The two main figures of merit of power devices are their on resistance and breakdown voltage. The design of these devices is complicated by the trade-off between the requirements for minimum on resistance and maximum breakdown voltage. The difference in lateral diffusion determines the channel length of the device, which is very small and results in low on resistance. The n-type drift region increases the breakdown voltage. Figure 10.1 shows the evolution of an LDMOS from an n-MOSFET.

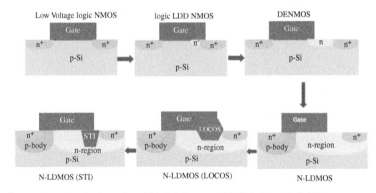

Figure 10.1 Evolution of an LDMOS from an NMOS. After Ref. [1].

Development of optimal LDMOS devices is a complex process, where several device design parameters need to be taken into account, and simultaneous optimization is needed towards maximum breakdown at minimal R_{on}, minimum area, high reliability performance, and ideal output characteristics. The 2005 International Technology Roadmap for Semiconductors states that the use of technology computer-aided design (TCAD) will provide as much as a 40% reduction in technology development costs. In today's competitive environment, time to market for new products is of paramount importance. Therefore, a methodology/ strategy is required that reduces the development cycle time. TCAD has become one of the key approaches to predict device performance and to optimize process conditions for CMOS development, as it is essential to debug process technologies before manufacturing release. Utilizing TCAD and design for manufacturability techniques, it is possible to reduce the total cycle time needed in the development of a technology.

In the following, a simple and predictive TCAD calibration methodology, which employs inverse modeling to support and enhance process technology development, is employed for the LDMOS process. Also, a detailed study for the LDMOS process development has been done via calibration and verification of the TCAD tools. Extensive use of design-of-experiment (DOE) matrices, based on TCAD simulations, is done. The results obtained are explained for the observed features. Layout parameters are varied and the electrical characteristics of the device (e.g., breakdown voltages, specific on resistance, etc.), together with hot-carrier behavior, are studied. DOE is used to determine the optimum process flow in order to achieve the desired electrical performance and analysis of the predicted breakdown distribution.

Sentaurus SProcess/SDevice simulators are used for process development and extracting the simulated electrical parameters. The process flow was developed using Sprocess, and the structure file was used as input into the device simulator SDevice. The initial process specifications (main process steps) are shown below. The process flow in this work represents a generic LDMOS technology. The major process steps are:

1. Substrate layer implant and epilayer growth
2. Well implants
3. Well drive-in
4. Local oxidation of silicon (LOCOS) oxide growth
5. Post-LOCOS cleanup
6. Gate oxide growth
7. Polysilicon gate deposition and reoxidation
8. Source–drain implants
9. Source–drain annealing
10. Body contact implants
11. Body contact annealing
12. Metallization

The simulations were based on models used for the smart power process from which the flow under investigation was derived, therefore providing some confidence in the usefulness of the models, and hence, TCAD tools used. In process simulations a fixed-meshing strategy as well as an adaptive-meshing strategy was employed. The Sentaurus Device performs device simulations to extract key electrical parameters in order to facilitate customized calibration and optimization. The process model was calibrated by including as much process information in the model as possible.

The process flow was simulated with Sentaurus Process using the equilibrium diffusion model and the parameters are set to the TSUPREM-4 defaults values. The two-phase segregation model is activated during LOCOS growth and segregation was neglected. The advantage is that the number of simulation runs to explore the complete parameter space and the resulting device electrical properties can be used to help optimize the process control settings and the analysis of the impact of process parameter sensitivity on device response provides greater insight into, for example, which process parameter would require monitoring inline. Figure 10.2 shows a schematic cross section of a typical LDMOS device.

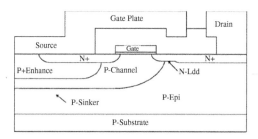

Figure 10.2 Schematic cross section of a typical LDMOS device.

In the simulation, 2.5 nm thick gate oxides were used. The n-well/p-well regions were simulated in 0.3 μm CMOS technology. The whole structure is isolated through the use of an n-well/n-tub/n-type buried layer. After each processing step, the simulated intermediate structure is saved. The subsequent processing steps are performed in a new simulation and the previous simulation results are reloaded. Major process simulation–generated structure files, that is, the process steps, are shown in Fig. 10.3a–f. Figures

10.3a and 10.3b show the close-up of the n-well and p-well corner after the well drive-in processing step. It has been observed that adaptive meshing results in a good resolution at the p–n junction. A similar result can be obtained with a fixed mesh.

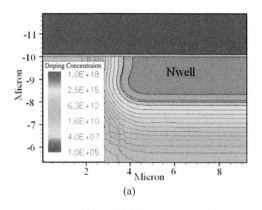

(a)

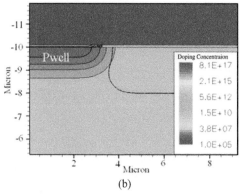

(b)

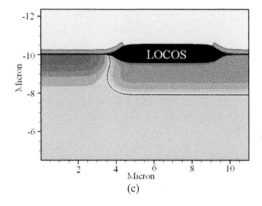

(c)

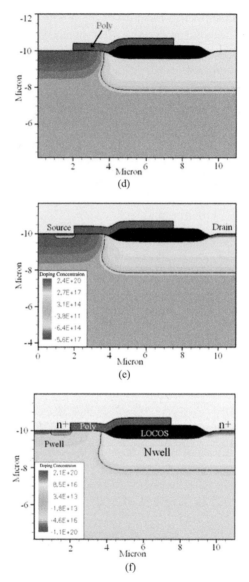

Figure 10.3 Simulation of an LDMOS process close-up of (a) an n-well corner and (b) a p-well corner after well drive-in. (c) Simulation of an LDMOS process close-up of a drain-side LOCOS edge after oxidation. (d) Simulation of an LDMOS process polysilicon gate deposition and reoxidation. (e) Simulation of LDMOS process source–drain annealing: concentrations of dopants in various regions are shown. (f) Simulation of LDMOS process body contact implant and annealing.

Figure 10.3c shows a close-up of the drain-side LOCOS edge after the oxidation step. It shows the mesh created by the fixed-meshing strategy and the adaptive-meshing strategy. In the LOCOS process definition, the temperature is first ramped from 700°C to 1000°C in an inert environment. Then, the structure is exposed to an oxidizing environment for 3 hours. Finally, the temperature is ramped down in an inert environment. As previously discussed, adaptive remeshing is active only during the inert diffusion parts. Figure 10.3d shows the polysilicon gate deposition and reoxidation. Figure 10.3e shows the source–drain annealing and concentrations of dopants in various regions directly after the implantation. Figure 10.3f shows the structure after body contact implant and annealing. Figure 10.4 shows the final device structure at end of process simulation. Metal regions are gray, oxide regions are brown, and concentrations of dopants are shown in silicon body region and polysilicon gate.

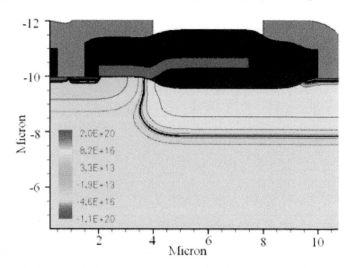

Figure 10.4 Final device structure at the end of the process simulation.

After the process simulation, device simulations were performed for I_d–V_g characteristics for a low drain bias, I_d–V_d characteristics for two different gate biases, and an off-state breakdown simulation. Sentaurus Device is used to simulate the drain current. In addition, the off-state breakdown characteristics are simulated. All *I–V* simulations are performed using a two-carrier drift-diffusion transport model. For the I_d–V_d simulations, lattice self-heating effects are included by solving the lattice temperature equation.

For all simulations, the doping-dependent mobility model is used, and high-field saturation effects and mobility degradation at the silicon–oxide interface are accounted for. The Shockley–Read–Hall (SRH) and Auger recombination models are included. For the breakdown simulation, the avalanche impact ionization model is activated. Figures 10.5 and 10.6 show the I_d–V_g and the I_d–V_d characteristics, respectively. Figure 10.7 shows the off-state breakdown characteristics for the LDMOS structure simulated with a fixed mesh and with adaptive meshing. It can be seen that both meshing strategies give virtually the same results.

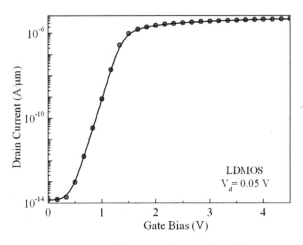

Figure 10.5 Drain current as a function of gate voltage for LDMOS structures simulated with a fixed-meshing strategy and an adaptive-meshing strategy.

In addition, the capacitance–voltage (C–V) simulation can provide detailed information about the channel region. Figure 10.8 shows the drain and source grounded C–V characteristics. At s negative gate voltage, the channel is in accumulation, and with increasing V_g, the drain side of the device enters depletion, the source side is in accumulation, and the center of the channel is at the flat band. As the gate voltage continues to increase, the depletion layer reduces the capacitance until around $V_g = 0$ V. At this point, the drain side of the device enters inversion.

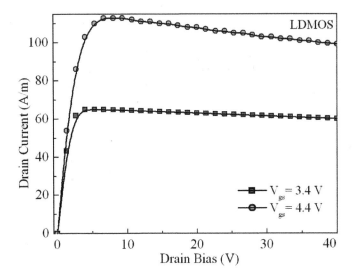

Figure 10.6 Drain current as a function of drain voltage for LDMOS structures simulated with a fixed-meshing strategy and an adaptive-meshing strategy for gate biases of 3.4 V and 4.4 V.

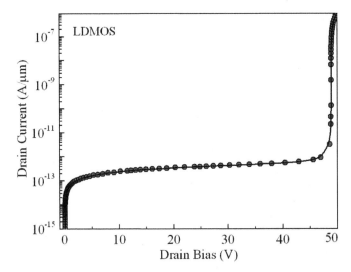

Figure 10.7 Off-state breakdown: drain current as a function of drain voltage for LDMOS structures simulated with a fixed-meshing strategy and an adaptive-meshing strategy for a gate bias of 0 V.

Currently, substantial research is being carried out to validate CMOS processes for radio-frequency (RF) applications. Many of these attempts are concerned with efforts to improve the device structure to optimize key performance measures such as unity gain cutoff frequency f_T, maximum oscillation frequency f_{max}. S-parameters provide an evaluation of the high-frequency performance. S-parameters have been simulated for frequencies in the range of 30–40 MHz. Gate-to-gate (C_{gg}), gate-to-source (C_{gs}), and gate-to-drain (C_{gd}) capacitances and transconductance have been deduced from Y-parameters, as follows:

$$C_{gg} = \frac{\text{Im}(Y_{11})}{\omega} \qquad C_{gs} + C_{gb} = \frac{1}{\omega}\text{Im}(Y_{11} + Y_{12})$$

$$C_{gd} = -\frac{\text{Im}(Y_{12})}{\omega} \qquad G_M = \text{Re}(Y_{21} - Y_{12})$$

The simulated S-parameters are shown in Fig. 10.9 and the extracted f_T and f_{max} are shown in Fig. 10.10.

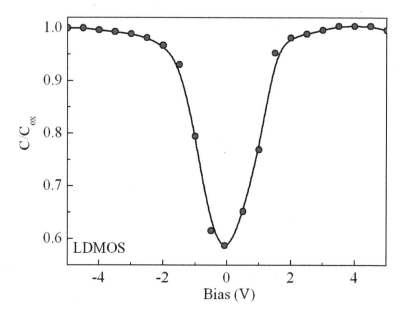

Figure 10.8 Simulated C_{gg} for analysis of the channel region.

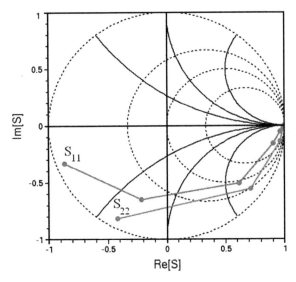

Figure 10.9 Simulated S-parameters for the LDMOS structure.

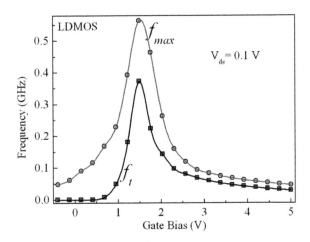

Figure 10.10 Simulated f_T and f_{max} for the LDMOS structure.

10.2 Vertically Diffused MOS Devices

Power MOSFETs are used as high current and voltage drivers in automotive, telecommunication, and power industries. With their superior switching speed, power MOSFETs have become one of the well-known technologies used in high-power systems. Solid-state power devices are capable of handling a wide range of currents and voltages. Although solid-state power MOSFETs utilize semiconductor processing techniques that are comparable to those of conventional very-large-scale integration (VLSI) circuits, voltage and current levels are much different from the design used in VLSI devices. A high blocking voltage in the off state and a high current capacity in the on state are the two important characteristics associated with power devices and usually exceed the limitations of modern conventional transistors to perform as a high-power switch.

MOSFETs for power management applications have a wide range of maximum voltage ratings from 10 to 1500 volts. Low voltage ratings (<30 V) are typically used for power switches in portable electronic equipment. While the basic physics of conventional integrated circuit devices is applicable, additional issues need to be considered when high-power BJTs and MOSFETs are used in power applications. For power applications, MOS transistors need a reasonable, short channel length and a low doping level in the drain region. Next, we discuss the development, fabrication, and characterization of discrete lateral-diffused vertical drain metal–oxide–semiconductor (VDMOS) power transistors.

Typical power VDMOS transistor fabrication steps were used to create a 50 V power VDMOS transistor structure were designed using Silvaco ATHENA process simulation. Modifications were made to include the Faraday shield and unclamped inductive switching (UIS) implant steps to address certain parasitic effects. UIS is used to reduce the parasitic bipolar junction transistor (BJT) operation or avalanche failure. The Faraday shield implant is performed to shift the parasitic gate-field capacitance. The ATHENA process simulation was used for the process simulation of the fabrication processes. The ATLAS device simulator was used for the device simulation. The combination of both made it possible to determine the impact of process parameters on device characteristics.

The fully coupled diffusion model (full.cpl) was used to account for the interaction between the impurity atoms and point defects, thus accounting for interstitials that were created during oxidation. A

VDMOS structure with parameters given in Table 10.1 was simulated both with and without a UIS implant to ensure this implant step would truly reduce the effect of a parasitic BJT. Figure 10.11 shows a 2D profile of a VDMOS structure showing three different terminals and different diffusion regions simulated using ATHENA.

Table 10.1 Parameters used for the VDMOS process design

Substrate thickness (µm): 650 to 700
Epilayer thickness (µm): 20 to 25
Epilayer resistivity (Ω-cm): 3 to 5
Gate oxide thickness (nm): 100
Polysilicon gate thickness (µm): 0.6
Body dose (cm^{-2}): 5×10^{13}
Body implant energy (keV): 100
Body drive-in time (min): 480
Body drive-in temperature (°C): 1100
UIS dose (cm^{-2}): 1×10^{14}
UIS implant energy (keV): 300
Source dose (cm^{-2}): 4×10^{15}
Source implant energy (keV): 65
Source/UIS anneal time (min): 20
Source drive-in temperature (°C): 1050

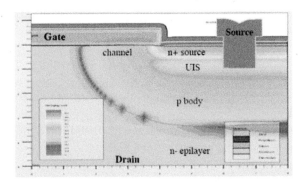

Figure 10.11 Two-dimensional profile of a VDMOS structure showing three different terminals and different diffusion regions simulated using ATHENA. After Ref. [2].

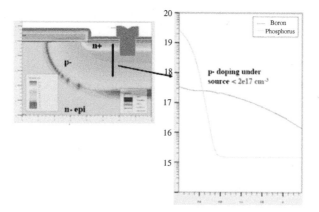

Figure 10.12 ATHENA simulation result showing the p-doping concentration under the source region for the structure without a UIS implant step. After Ref. [2].

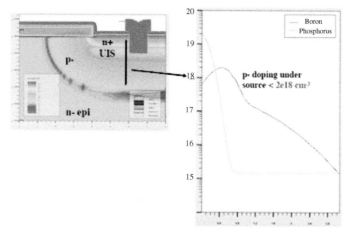

Figure 10.13 ATHENA simulation results showing the p-doping concentration under the source region for the structure with a UIS implant step. After Ref. [2].

Figures 10.12 and 10.13 illustrate ATHENA simulation results showing the p-doping concentration under the source region of a structure with and without a UIS implant step. The structure with the UIS implant has a boron concentration an order of magnitude higher than that of the structure without the UIS implant. This higher doped region should reduce the base resistance value, while increasing the doping in the parasitic base and eliminating the current gain mechanism. Figure 10.14 shows I_d–V_g characteristic characteristics of the simulated VDMOS structures with and without the UIS implant, verifying that the additional UIS implant did not have any significant effect on V_t or I_d behavior. Figure 10.15 shows the family of characteristics of the simulated VDMOS structure with the UIS implant. The two most important DC characteristics of a device (the maximum drain current and the threshold voltage) did not change with the addition of the UIS implant step.

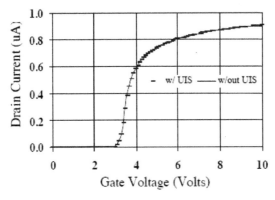

Figure 10.14 I_d–V_g characteristic characteristics of simulated VDMOS structures with and without a UIS implant.

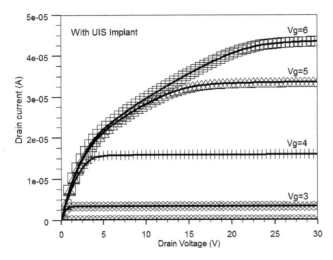

Figure 10.15 Family of characteristics of a simulated VDMOS structure with a UIS implant.

10.3 Summary

The strategy utilized in this chapter was to use TCAD where process and device simulation is used in conjunction with DOEs to reduce the process development time. The DOE approach based on TCAD simulation has been used for the design of a medium-voltage (40–50 V) LDMOS. Lateral-diffused VDMOS power transistors using interdigitated source/gate design have also been considered. The TCAD methodology can be expanded to more critical process design, analysis, and ultimately support of shorter design cycles.

References

1. Aghoram, U. (2010). *Impact of Strain on Memory and Lateral Power MOSFETs*, PhD thesis, University of Florida.

2. Tokunaga, K. (2008). *Development, Fabrication, and Characterization of a Vertical-Diffused MOS Process for Power RF Applications*, MS thesis, Rochester Institute of Technology.

Chapter 11

Solar Cells

The increasing demand for energy over the past 50 years has
resulted in a drive for research into possible alternative energy
sources that are both commercially viable and do not create waste
products that are detrimental to the environment. It has been
projected that by 2050, the worldwide energy consumption will
increase to approximately 28 TW, from its 2006 level of around
11 TW. Silicon-based cells have dominated the photovoltaic market
for the last 50 years and much research is still made into enhancing
existing systems. However, silicon cells have a theoretical maximum
power conversion capacity of less than 29%. Alternate forms of
cells are thus being heavily researched, particularly those using
nanomaterial, due to the favorably small amount of material needed
and the likely associated decrease in cost per watt. Current silicon-
based solar cells on the market, based on inorganic solid-state
junction devices, are environmentally clean, but expensive due to
large amounts of materials required for production, and generally
heavy or cumbersome. Though silicon cells have relatively high
reported power conversion efficiencies of between 15% and 20%,
the efficiencies achieved in laboratory studies are often not conveyed
in commercially available applications, due to problems with scale-
up and the requirement of highly controlled conditions.

As continuous innovation makes solar cells more complex
involving more process and geometrical variables, it is impractical

Introducing Technology Computer-Aided Design (TCAD): Fundamentals, Simulations, and Applications
C. K. Maiti
Copyright © 2017 Pan Stanford Publishing Pte. Ltd.
ISBN 978-981-4745-51-2 (Hardcover), 978-1-315-36450-6 (eBook)
www.panstanford.com

to design new cells without simulation as too many experiments are needed to investigate design space and risks missing optimum design. To use solar energy in an efficient way, it is necessary to understand the nature of the spectrum. The solar radiation spectra show the incoming levels of energy at different wavelengths. The process of absorption of light (photon) is one of the principal aspects of electricity conversion. For example, the most commonly used semiconductor is silicon, having a bandgap of 1.1 eV, which covers most of the radiation spectra. For higher efficiency, collection of as much charge as possible, increase in the depth of the thick active layers is essential, which, however, results in high recombination rates before collection. To avoid this problem, the thickness of the p–n junction needs to be kept to a reasonable minimum, compromising charge generation over collection.

Solar cells to systems involve design and simulation at three levels, viz., at the cell level, at the module level, and at the system level. At the cell level attention is given to optimize geometric and process parameters to maximize the efficiency. At the module level one needs to study the effects of interconnects on performance, maximum power point as a function of luminance and cell width and minimizes the impact of cell variation or degradation on module performance. At the system level it is necessary to maximize system performance accounting for diurnal solar inclination and tracking of solar path and system level efficiency in relation to power delivered to the grid, including the inverter system. Solar cell optimization involves several steps such as selection of parameters to be investigated, unit cell pitch, active layer thickness, doping, and lifetime, and surface recombination velocity shows a major influence on cell response. Back-contact optimization involves metal finger pitch to achieve good performance with low-cost screen-printing manufacturing and simulation to correctly capture the measured behavior across a range of contact pitch and bulk resistivities. Once cell behavior is optimized, one needs the system level simulation to (a) minimize interconnect losses and (b) evaluate effects of environmental variation such as light intensity and incidence angle and temperature variation.

The efficiency of solar cells depends on:

- The spectral absorbance of the light-absorbing material and the electron/hole injection efficiency

- The electron transport and recombination rates, which are dependent on the electron-conducting material
- The hole transport and stability, which is related to the hole conductor used

11.1 Solar Cell Simulation

In the following example, we simulate a solar cell structure using ATHENA and ATLAS. The input file consists of the following steps for simulation:

- Construction of solar cell doping and geometry in ATHENA
- Simulation of short-circuit current
- Simulation of open-circuit voltage
- Simulation of spectral response
- Simulation of illuminated and dark I–V characteristics

ATHENA is used to create a typical solar cell structure. The device used is a diode of n^+ over a p-substrate. The junction depth is approximately 0.25 µm. A single contact is placed in the center of the structure. The fabrication process consists of an implant and diffusion followed by electrode formation. Figure 11.1 shows the solar cell structure in which the photogeneration rate is shown. The short-circuit current is the current when anode and cathode are shorted. This is simulated by illuminating the device with zero voltage on all contacts. It is possible to store the optical intensity of the illumination by specifying output opt.int at any time before saving a structure file. The photogeneration rate will appear in the solution structure file by default. The structure stored at this point can be plotted to show contours of photogeneration, carrier concentration or potential. The photogeneration contours show the effect of the opaque cathode contact. At the start the definition of material parameters and opaque metal contact is repeated. For this test a different light beam is required. A beam origin and angle of incidence are set as before. The short-circuit case is considered so zero biases are set on both electrodes. The wavelength parameter λ of the solve statement is used to set the wavelength of the incident light. The range of wavelength used was 300 nm to 1000 nm.

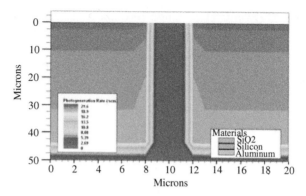

Figure 11.1 Solar cell structure in which the photogeneration rate is shown.

Plotting the resulting log file it is possible to see how the cathode current varies with wavelength. Setting a plot versus wavelength of the source photocurrent (current available in the light beam), the available photocurrent (current available for collection) and the actual cathode current can show how the device behaves (Fig. 11.2). The losses between source and available photocurrent are caused by reflections and transmission. This dominates the losses at all except the shortest wavelengths. The losses from available photocurrent to the actual simulated cathode current are due to recombination. These are very low except at the extremes.

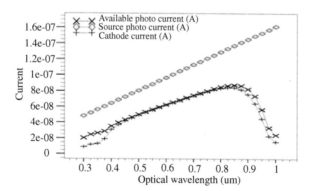

Figure 11.2 Cathode current variation with wavelength.

The plot of illuminated and dark *I–V* characteristics, allowing the designer to choose a suitable load line for the device under normal operating conditions is shown in Fig. 11.3. A number of useful parameters are can be extracted from this characteristics using the extract statement and a plot of efficiency versus light frequency.

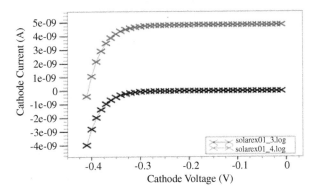

Figure 11.3 Illuminated and dark *I–V* characteristics of a solar cell.

11.2 Organic Solar Cells

In recent times there has been considerable interest in the use of organic materials in display technologies. Organic solar cells have become an important field of study in recent years. The simplicity of processing, the low temperature required for fabrication, and availability make them very attractive from a cost point of view. Recognition of the benefits of inexpensive materials and processing as well as large-area conformal construction have also created high interest in organic solutions for solar energy applications. In particular, conjugated polymer solar cells are among the simplest and cheaper of all. Since their first formal appearance in 1987, now with different configurations, their efficiency has slowly increased. Organic solar cells differ greatly in composition from inorganic cells, but the physics governing the processes are almost similar. Whereas in inorganic cells the active region of the cell is generally a lattice of semiconductor material, organic cells have active regions composed of complex organic molecules in solution.

Organic cells utilize a bilayer heterojunction structure, where there would be two distinct layers of organic materials. One layer would be composed of an electron-accepting material and the other of an electron-donating material, very much akin to p-type and n-type materials. The important distinction to make between inorganic cells and organic cells is that in inorganic cells, photons stimulate electrons into the conduction band, directly creating electron–hole pairs. Once excited, the carriers are able to move through the conduction band to elsewhere in the lattice. In organic cells, carrier generation requires both a donor and an acceptor material. This process starts with a donor molecule absorbing a photon, generating an excited electron. This electron is still bound to the donor material, so another step is required. If the donor material is close enough to an acceptor material, again a very small distance since the mobility of the organic materials is so low, the excited electron can move to and excited state in the acceptor molecule, generating a carrier pair. Since the generation process requires both donor and acceptor molecules, the region they are generated in is limited to the thin region near the interface of the two layers. Carrier lifetime is significantly improved in this design compared to the single layer design as once carriers enter the donor/acceptor regions recombination is unlikely, dramatically improving efficiency. The organic material absorbs the incident photons and generates carriers. Since the diffusion length for electrons is short in organic films, on the order of only 20 nm, these cells exhibit very low conversion efficiencies, on the order of 0.01% or lower.

One of the biggest problems in organic materials is charge collection. Due to their nature, the hole–electron pairs generated have bigger bonding energies than in inorganic materials. Today polymer solar cells are based on the bulk heterojunction (BHJ) concept and to increase efficiency further, tandem configurations are being developed. Normally, the bandgap of conjugated polymers is around 2 eV and it is clear that this type of bandgap would leave almost half of photons being unabsorbed. Low-bandgap materials, reaching a bandgap of 1.2 eV, have been processed giving some hope to the solar cell development area. Attempts for tandem solar cells including low-bandgap materials are being made.

Next, the modeling and design of a polymer BHJ solar cell will be presented. The device, comprised of layers of BHJs, takes advantage of the diverse absorption characteristics of the materials involved to improve efficiency. This approach takes into consideration Langevin recombination constants and quantum efficiency modifications to fit existing measured data. The characterized layers are then used to simulate the tandem solar cell response to the solar spectra.

Silvaco has long been a leading supplier of organic solar cell simulator which will be used in the simulation. Organic solar models for BHJ organic solar cells will be used. In this model, the BHJ solar cells consist of an interpenetrating mixture of donor and acceptor materials. Absorbed light generates excess exciton which is charge neutral and diffuses to the acceptor–donor interface where they dissociate into free electron–hole pairs. The electron–hole pairs are then separated by the internal field and are swept up at the contacts as in crystalline photodetectors. The model includes the standard drift-diffusion equations (Poisson's equation and the electron and hole continuity equations) augmented by the singlet exciton continuity equation.

In this case study we use singlet exciton dissociation in the modeling of organic/polymer solar cell devices. As the modeling is based on the BHJ concept, we do not consider the detail of the location of the heterointerface as it is considered as a random interpenetrating mixture of materials. It is assumed that the material is electrically uniform and that the mean distance from anywhere to the heterointerface. In the simulation, the base material considered for the active layers is organic. This material belongs to the Organics group in the Silvaco material library. There are three keys for simulation of exciton dissociation. The specific models that are included in the simulation are exciton generation, dissociation, and molecular recombination characteristics. They are the inclusion of the singlet exciton continuity equation, the inclusion of Langevin recombination and the inclusion of the singlet dissociation model. There are two main aspects of the material specification related to the simulation, optical and electrical properties. Most of these parameters (Tables 11.1 and 11.2) are associated with the continuity equations of the system governing the electrical behavior of the material.

Table 11.1 List of the parameters used to characterize the materials involved

nv300	Valence band density at 300 K	cm^{-3}
μ_n	Low-field electron mobility	cm^2/V.s
μ_p	Low-field hole mobility	cm^2/V.s
τ_{n0}	Shockley–Read–Hall (SRH) lifetime for electrons	s
τ_{p0}	SRH lifetime for holes	s
h_a	Total acceptor-like trap density	cm^{-3}
h_d	Total donor-like trap density	cm^{-3}
s.binding	Singlet binding energy	eV

Table 11.2 Common parameters used for all the bulk heterojunction layers

Parameter	Value	Source
Affinity	3.75 eV	–
nc300	2.8 × 10^{19} cm^{-3}	Silvaco
nv300	1.04 × 10^{19} cm^{-3}	Silvaco
s.binding	0.4 eV	Fixed-mean
rst.exciton	0.25	Silvaco
lds.exciton	0.01 microns	Silvaco
qe.exciton	1	Silvaco
a.singlet	1.2 nm	–
a.langevin	1	Silvaco

As part of the models, the source of energy considered for this simulation needs to be specified. In this case the source of light is the solar spectra AM1.5. To include this information in the BEAM statement, the data was processed and a file was created according to Silvaco's specification of power density versus frequency. After all the materials parameters have been specified, the models need to be introduced. The three models are enabled on the MODEL statement by the parameters: SINGLET, LANGEVIN and S.DISSOC. QE.EXCITON parameter of the MATERIAL statement is used which characterizes the number of singlet exciton generated for each photon that is absorbed. Absorbed photons not generating singlet generate electron–hole pairs. The main interest in the device characterization and simulation is to obtain the efficiency of the cells. For this, the SOLVE statement is set up to change the voltage of the anode in increments of 0.05 V. The overall power of the light source needs to be specified here as well. During simulation, when the processing reaches this point, Silvaco calls the Organic Solar module using the power spectra specified in the BEAM statement. Figure 11.4 shows the solar cell *I–V* characteristics.

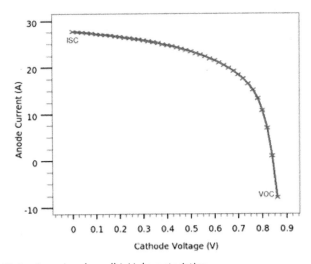

Figure 11.4   Organic solar cell *I–V* characteristics.

11.3   Tandem Solar Cells

The first homojunction silicon solar cell developed during 1954 had an efficiency of only 6%. But the progress in silicon technology resulted in single-crystal silicon solar cells with higher reliability and efficiencies reaching above 25%. The silicon photovoltaic modules are very expensive. However, about 95% of commercially available photovoltaic modules are made from crystalline silicon cells and only 5% is contributed by thin-film technology. Thin-film solar cells have become a strong competitor for the single-crystal and polycrystalline silicon solar cells because of the cheaper raw material as well as processing costs. Substrates used in thin-film solar cells are a metal or a glass or a polymer. These are relatively cheaper compared to silicon wafers. A tandem solar cell usually has two cells having different bandgaps placed one over other with an encapsulated between them. They are designed in such a way that the top cell has a larger bandgap and hence can absorb shorter wavelengths. The bottom cell with a relatively smaller bandgap can absorb longer wavelengths that pass through the top cell. To obtain efficiencies greater than 25%, the top cell should have a bandgap of 1.7 eV. This would mean that the top cell has to contribute to two-thirds of the total efficiency. The bottom cell of the tandem structure has a bandgap of 1 eV. Tandem solar cells fabricated from thin films provide promise of improved efficiency, while keeping the processing costs low.

Amorphous silicon does not require a costly a manufacturing process, which reduces the cost of the Si-based solar cells. Process temperatures tend to be significantly lower than in production of crystalline Si, allowing the use of much less expensive substrates, including float glass or foils made of metal or plastic if amorphous silicon is used. Additionally, the absorption coefficient of amorphous silicon is significantly higher (upward of 20x) than that of crystalline silicon, with layers on the order of microns able to absorb nearly all incident light on its surface. Carrier generation occurs much like in crystalline silicon, with one noteworthy exception. Despite still being composed mainly of silicon (hydrogen is often present in the material as well, deposited during the production process), the effective bandgap of amorphous silicon can range between ~1.55–1.8 eV, compared to 1.1 eV for crystalline silicon. Not only is the bandgap significantly higher than crystalline silicon, there is a certain amount of flexibility that can be used to tailor the material to the needed application. These factors make amorphous silicon a cheap and versatile photovoltaic device. However, degradation is also a significant issue in amorphous silicon cells. When exposed to sunlight, a phenomenon known as the Staebler–Wronski effect occurs in amorphous photovoltaic. Upon exposure to sunlight, even as briefly as a few hours, photoconductivity of the material decreases.

In the following, we consider the design and simulation of a-Si:H/μc-Si:H tandem solar cell (from Silvaco) with the following structure in the simulation: ZnO:Al(500nm)/p-a-Si:H(10nm)/i-a-Si:H(200nm)/n-a-Si:H(15nm)/p-μc-Si:H(10nm)/i-μc-Si:H(2.2μm)/

p-μc-Si:H(15nm)/ZnO(100nm)/Ag. Both *I–V* and external quantum efficiency (EQE) simulations of a tandem a-Si:H/μc-Si:H solar cell are performed.

In simulation, the Tauc–Lorentz dielectric function with the Urbach tail model for a complex index of refraction was used for a-Si:H material. The model parameters were calibrated to fit the EQE of the top cell. For the bottom cell silicon index of refraction was used except that the imaginary part of the index of refraction was shifted in energy to take into account the bandgap difference between silicon and μc-Si:H material. To model the interlayer between the two solar cells a simple method was used. The method used here consists of adding an electrode which exactly overlays the interlayer and attaching a lumped resistance to it using the contact name=com resist=1e16 statement. In doing this, we force the current to flow from the anode to the cathode and prevent any current to flow in the added electrode. Physically it can be justified by the fact the interlayer is acting like a resistor letting current flows without significant limitation. The value of the resistance can be used to adjust the amount of current flowing through the added electrode, thus controlling the interlayer resistance. In hydrogenated amorphous silicon (a-Si:H), the effect of dangling-bond states on recombination can be significant. The dangling-bond states are amphoteric and located around the middle of the bandgap. In this example the **amphoteric** parameter on the **defect** statement was used to specify amphoteric defects. Figures 11.5 and 11.6 show the *I–V* characteristics and the EQE simulations of a tandem a-Si:H/μc-Si:H solar cell, respectively.

Thin-film solar cell modules based on copper indium gallium selenide (CIGS) are a technology with great potential. Two reasons for this are low material consumption and relatively high efficiency. The following example demonstrates CIGS solar cell module simulation and optimization. In a CIGS solar cell module several cells are connected in series. The cells are the active area of the module whereas the interconnections between cells are the dead area. The efficiency of the module is thus dependent on the number of cells in the module. Thus there is a trade-off between maximizing absorption and minimizing resistance of the interconnection. Figure 11.7 shows the efficiency dependence on the number of cells in a CIGS solar module.

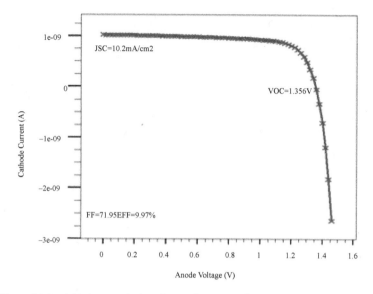

Figure 11.5 *I–V* characteristics of a tandem a-Si:H/μc-Si:H solar cell.

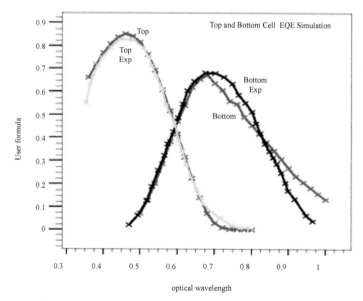

Figure 11.6 EQE simulations of a tandem a-Si:H/μc-Si:H solar cell.

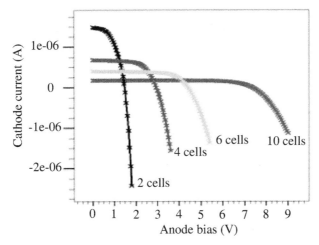

Figure 11.7 Efficiency dependence on the number of cells in a CIGS solar module.

11.4   3D Solar Cell Simulation

In the following case study we simulate a 3D solar cell with and without lenses. The simulations performed are:

- 3D process simulation of the solar cell using VictoryCell
- Mesh generation of the 3D solar cell structure
- 3D device simulation of I–V characteristics using VictoryDevice with and without lenses

The 3D process simulation starts from a mask and consist of a deposition and angled etch of an oxide layer on top of silicon in order to define oxide lenses. The shape of the lenses can be adjusted as a function of mask size and angle used during the etch. At the end of the process simulation a 3D structure is saved and meshed using a full 3D Delaunay mesh. Three-dimensional structures are shown with the lens after process simulation (Fig. 11.8) and with rays (Fig. 11.9), respectively. 3D refinement of the mesh is done as a function of net doping. The 3D structure is then passed to VictoryDevice, where optical and electro-optical simulations are performed. Three-dimensional ray tracing is used during the simulation. This is a method by which beams of light are traced through a structure, taking into account reflection, refraction, and attenuation. The calculated light intensities and absorptions are then used to calculate the photogeneration rates are shown in Figs. 11.10 and 11.11, respectively. Whenever a ray encounters a region interface, or the incident ray encounters a device boundary, the ray is split into a reflected and a transmitted beam. For complex structures (especially in 3D) this can result in a large number of rays. The ANGLE (or PHI) parameter specifies the direction of propagation of the beam relative to the *x* axis. In 3D, one can specify the angle THETA, which is the rotation about the *y* axis. By default the beam will automatically be sized to illuminate the entire device. The optical and electrical simulations are done using an iterative multithreading domain decomposition–based solver called PAM. MPI. Figure 11.12 shows the *I–V* characteristics of the solar cell with and without a lens.

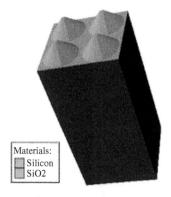

Figure 11.8 3D structure of a solar cell after process simulation with a lens.

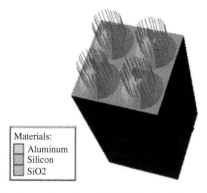

Figure 11.9 3D structure of a solar cell with illumination on the lens.

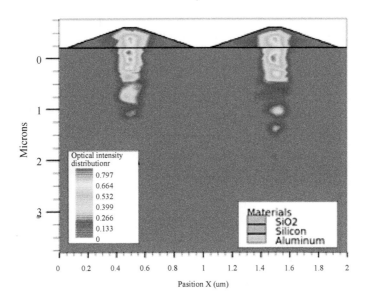

Figure 11.10 Optical intensity distribution in a solar cell.

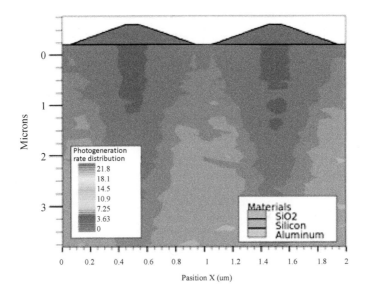

Figure 11.11 Photogeneration rate distribution in a solar cell.

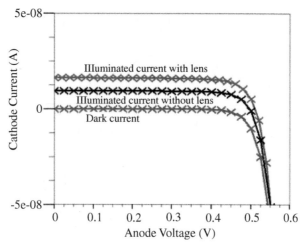

Figure 11.12 *I–V* characteristics with and without a lens.

11.5 Summary

The objective of this chapter has been to design different types of solar cells. Silvaco TCAD tools provide a solution for researchers interested in solar cell technology. They enable researchers to study the electrical properties of solar cells under illumination in both 2D and 3D domains. The simulated properties include *I–V* characteristics, spectral response, quantum efficiency, photogeneration rates, potential distribution, etc. In this chapter, three types of solar cells, viz., organic solar cells, tandem solar cells, and 3D solar cells, have been investigated.

Chapter 12

TCAD for SPICE Parameter Extraction

Technology scaling is increasing the complexity and nonideality of the electrical behavior of semiconductor devices. Besides providing a deep insight, especially for aggressively scaled devices, TCAD simulations may help in the generation of predictive models which play a crucial role in reducing circuit development cycle times and costs in semiconductor industry. During the last two decades circuit simulation has become an indispensable tool for the design of integrated circuits and compact device model is an essential part of the simulation (see Ref. [5] of Chapter 1). Device modeling also plays an important role in semiconductor fabrication, especially for the growing diversity of device requirements. In this chapter we review the various models currently in use in the semiconductor industry. We discuss on device characterization using available semiconductor device analyzer equipment, parameter extraction software, and methodologies used today.

12.1 Compact Model Generation

During the last four decades circuit simulation has become an indispensable tool for the design of integrated circuits (ICs) and device models are an essential part of this simulation. The traditional link between design and technology has been defined mainly via the

Introducing Technology Computer-Aided Design (TCAD): Fundamentals, Simulations, and Applications
C. K. Maiti
Copyright © 2017 Pan Stanford Publishing Pte. Ltd.
ISBN 978-981-4745-51-2 (Hardcover), 978-1-315-36450-6 (eBook)
www.panstanford.com

compact circuit modeling of devices. Compact device modeling may be seen as a bridge between the IC design and the chip fabrication. However, compact model development is becoming difficult to keep pace with the introduction of novel performance–boosting technologies. Circuit simulation needs compact device models. Requirements for the advanced compact model are illustrated in Fig. 12.1. Models of the performance enhancement techniques such as stress engineering, high-k dielectric, and thin-body multigate device must be implemented. New phenomena, including the layout proximity effect, quantum effect, ballistic transport, and enhanced overlap capacitance models, are required. Circuit designers cannot begin to design for a significant period of time before the processing, characterization, and design kit cycle is complete. Technology computer-aided design (TCAD) can be used to generate the SPICE model for a target technology under development for concurrent development. A typical model extraction procedure includes:

- Use of only device simulation to extract SPICE parameters:
  - Define test structures.
  - Generate I–V data using calibrated device models.
  - Extract SPICE model parameters.
- Process/device TCAD-based SPICE model parameter extraction:
  - Calibrate TCAD models for existing/mature technology.
  - Perform process simulation to optimize new process recipe/target.
  - Perform device simulation to optimize new device specification/target.
  - Generate I–V and C–V data.
  - Extract SPICE model using parameter extraction tool such as Paramos.

The philosophy behind the compact modeling is to come up with closed-form equations that represent the TCAD data with all the nonlinearities included and yet efficient to use. The following is a brief guideline for formulating compact models:

- The approach is different from the analytic formulation starting from physical assumptions, since it is assumed that the entire complex device physics has already been included in the TCAD data. The formulation should be based on empirical nonlinear regression to curve-fit the TCAD data.

- It is important to use physics-based equations whenever possible so that the functionality can represent the data in a wide range and different conditions. It is important to manipulate the data to find a "global" functionality.
- The parameters in a compact model should be unambiguously defined and should be easily extracted from the TCAD data. The extraction procedure should be easily automated since it is expected that it will be repeated if a new set of data is given.
- It is preferable that the compact model includes as many variables as possible such that different target–variable dependencies can be easily evaluated. However, there will be a trade-off in terms of the complexity in the formulation and extraction.
- Associated with each compact model, there should be an estimation of the relative error to the original TCAD data. In this way, the accuracy of the compact model can be quantified.
- In principle, the same procedure for the compact model formulation can be applied to the same set of data obtained from measurements, if available.

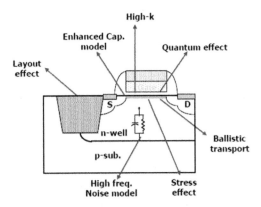

Figure 12.1 Requirements for the development of advanced compact models.

One can distinguish in principle three strategies for developing device models: physics-based, empirical, and tabular models. Physics-based models are based upon device physics. Parameters within these models are based upon physical properties such as oxide thicknesses, substrate doping concentrations, carrier mobility, etc. An empirical model is based upon curve fitting, using whatever functions and parameter values most adequately fit measured data to enable simulation of transistor operation. A tabular model is a form of a look-up table containing a large number of values for common device parameters such as drain current and device parasitic. Table models are easier to construct but require high memory space and are not very much in use.

Circuit simulation is usually based on SPICE compact models. These are high-level models that reproduce the device characteristics through a set of deterministic analytical equations based on device physics. To represent a particular technology, SPICE models use a set of parameters, whose number can vary, to fit the data coming from electrical measurements. In Fig. 12.2, a flowchart of a unified reference, temperature, and geometry parameter extraction procedure for the scalable model is shown. It can be implemented in parameter extraction software such as ICCAP or UTMOST. The essential feature of the parameter extraction procedure is a direct extraction of the geometry parameters from the measured electrical characteristics and the desired modeling tool such as the vertical bipolar intercompany (VBIC) model. The complete parameter extraction procedure is integrated in the single environment, which saves time in verification of the final scalable model results with measured data.

In principle, a TCAD synthesis approach can be applied to the prediction and characterization of semiconductor technologies and devices. The compact models, which are formulated on the basis of the TCAD data, could, in principle, be extracted from the measured data if the same set of the measurement data as TCAD is available. The approach makes use of the expertise of professional modeling efforts and provides a solution for nonexpert engineers so that they can concentrate on the design optimization and trade-off and do not need to care about all the pitfalls in the modeling. As the technology development progresses, updates to the compact models can be provided to ensure the designers will have minimal design impact from the initial models to the final hardware-based models. The following section describes the methodology of the TCAD process, device, and compact modeling strategy commonly employed in developing state-of-the-art technologies.

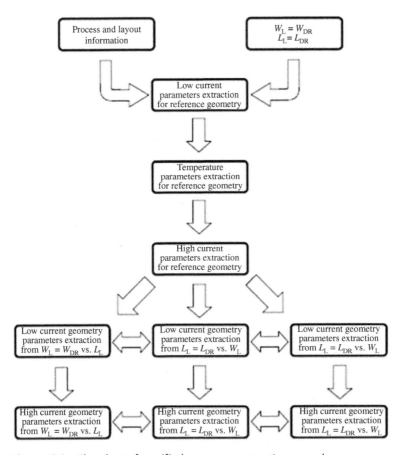

Figure 12.2 Flowchart of a unified parameter extraction procedure.

In fact, for chip design, realistic analog simulation models are necessary to simulate circuit performance with a sufficient degree of accuracy. The accuracy of circuit simulations not only depends on an accurate model but also on the used parameter extraction techniques. Compact model parameters are often obtained from extraction programs like ICCAP or UTMOST using measured devices over a wide range of lengths and widths. In the process of device compact modeling, the parameter extraction is very important. The purpose of the parameter extraction is to find values for model parameters and optimize the parameter values to achieve the best-possible fit between simulated and measured data for a given technology. In this chapter, automatic parameter extraction techniques of the

UTMOST extraction package are presented to extract compact model parameters.

12.2 Compact Modeling of HBTs

The first compact bipolar model was published in 1954 by Ebers and Moll; it is a very simple model but hardly used anymore for today's advanced analog circuit design. In 1970 Gummel and Poon published their integral charge control model, which was a big step forward: it included, among other features, high injection in the base. The SPICE version of the Gummel–Poon (SGP) model is still the workhorse of the bipolar designer today. It is well known that for bipolar compact modeling purpose, the Gummel–Poon (GP) model is not accurate for modeling the high-frequency operation of transistors. Accurate compact modeling requires the use of more complex models. The chosen models are the SGP model because it is widely known and can serve as a sort of reference here, the MEXTRAM model for its completeness and being used in industry for about 10 years, the HICUM model for its proven usefulness in high-speed circuits, and the VBIC95 model proposed by the US electronics industry.

In the modern semiconductor industry, compact models play an important role in connecting semiconductor manufacturing and circuit design. Successful IC design relies heavily on accurate SPICE modeling. The area of compact modeling for SiGe heterojunction bipolar transistors (HBTs) is, therefore, crucial to the successful introduction of the SiGe technology to the IC industry. At high frequencies, many parasitic effects come into play and make the physical extraction of device parameters difficult. Significant testing and calibration and de-embedding issues must be overcome for the proper modeling of RF devices. The testing issues may be more difficult for state-of-the-art SiGe HBTs because of their very high cutoff frequencies. HBT models should describe as accurately as possible the following effects:

- High injection in the base
- Quasi-saturation in epilayer, including the Kirk effect
- Avalanche multiplication
- Base resistance
- Noise; $1/f$, shot and thermal noise

- Self-heating
- Collector–base depletion capacitance

The SGP model has been in use for the last 30 years because it is convenient and computationally efficient and most people in the industry prefer to use it unless other more advanced models prove necessary. However, the intrinsic limitations of the SGP model at radio frequency (RF) are well known. In addition to the most popular SGP model, some other advanced models, such as high current bipolar compact transistor model (HICUM), MEXTRAM (developed by Philips Semiconductors), and the VBIC model are being currently used for modeling of advanced devices. The MEXTRAM model has been extended to high-speed small-signal applications in Si bipolar junction transistors (BJTs) as well as to SiGe HBTs.

VBIC is a BJT model that was developed as an industry standard, public domain replacement for the SGP model. It is based on the extension of the Kull et al. model. A comparison of the models is made next. The SGP model has so far been the workhorse of the bipolar circuit designer. Advanced compact models such as HICUM, MEXTRAM, and VBIC are applied to model SiGe HBTs. More recent developments are in the MEXTRAM model, largely depending on the modeling of the total stored charge. In the United States efforts are undertaken to come to new standard models for metal–oxide–semiconductor (MOS) and bipolar devices. The proposal for a new compact bipolar model called VBIC95 has recently been introduced.

12.2.1 VBIC

VBIC allows accurate simulation, thanks to the following improvements:

- Improved Early effect modeling
- Physical separation of I_c and I_b
- Improved HBT-modeling capability
- Improved depletion and diffusion capacitances
- Parasitic PNP
- Modified Kull quasi-saturation modeling
- Constant overlap capacitances
- Weak avalanche model
- Base–emitter breakdown

- Improved temperature modeling
- Self-heating
- Parasitic fixed (oxide) capacitance modeling

12.2.2 MEXTRAM

The MEXTRAM model is a strongly physics-based model, which has been developed to address many of the shortcomings of the GP model. These shortcomings include:

- Constant bias-independent Early voltages
- Poor modeling of substrate effects
- No modeling of avalanche effects
- Poor geometry and temperature scaling
- Inadequate modeling of high-frequency effects
- Weak quasi-saturation modeling

The effects modeled by MEXTRAM include:

- Variation in base resistance due to emitter current crowding and conductivity modulation
- Monotonic bias-dependent forward and reverse Early effects
- Built-in electric field in the base region
- Monotonic Early voltages and f_T behavior for lightly doped epilayer devices
- First-order approximation of high-frequency effects in the intrinsic base (high-frequency current crowding and excess phase shift)
- Physical formulation of collector epilayer resistance, including hot-carrier behavior and current spreading
- High injection effects in the base
- Substrate effects, including parasitic p-n-p
- Weak avalanche effects
- Hard and quasi-saturation

MEXTRAM provides several features that the GP model lacks:

- Evolution of a cutoff frequency with a smoothing parameter
- Bias-dependent Early effect
- High injection effects
- Ohmic resistance of the epilayer

- Velocity saturation effects on the resistance of the epilayer
- Hard and quasi-saturation (including the Kirk effect)
- Split base–collector and base–emitter depletion capacitance
- Substrate effects and parasitic p-n-p current crowding and conductivity modulation of the base resistance

MEXTRAM is a publicly available compact model for bipolar transistors. It has a number of improvements with respect to the well-known SGP model, which makes it more accurate, especially in analog, RF, and large-signal simulations. MEXTRAM covers several effects that are not included in, for example, the original GP model. These effects include:

- Temperature
- Charge storage
- Substrate
- Parasitic p-n-p
- High injection
- Built-in electric field in the base region
- Bias-dependent Early effect
- Low-level, nonideal base currents
- Hard and quasi-saturation
- Weak avalanche
- Hot-carrier effects in the collector epilayer
- Explicit modeling of inactive regions
- Split base–collector depletion capacitance
- Current crowding and conductivity modulation for base resistance
- First-order approximation of distributed high-frequency effects in the intrinsic base (high-frequency current crowding and excess phase shift)

12.2.3   HICUM

The HICUM model is particularly well suited for a high-frequency and high-current description. The key improvement with regard to the GP model is that HICUM describes the Kirk effect and quasi-saturation via a semiphysical model for the transit time, which is an indirectly measurable quantity through the transit frequency.

The most important model features of a HICUM physics-based equivalent circuit containing all relevant effects for present bipolar process technologies are:

- Emitter periphery effects
- Distributed external base–collector region
- Parasitic bias-independent base–emitter and base–collector capacitances
- (Weak) avalanche breakdown
- Substrate coupling network
- Parasitic substrate transistor
  The major effects taken into account by HICUM are:
- High-current effects
- Distributed high-frequency model for the external base–collector region
- Emitter periphery injection and associated charge storage
- Emitter current crowding
- Two- and three-dimensional collector current spreading
- Parasitic (bias-dependent) capacitances between base–emitter and base–collector terminals
- Vertical non-quasi-static effects for transfer current and minority charge
- Temperature dependence and self-heating
- Weak avalanche breakdown at the base–collector junction
- Tunneling in the base–emitter junction
- Parasitic substrate transistor
- Bandgap differences occurring in HBTs
- Lateral scalability

12.3 Device Characterization

On-wafer measurement is preferred for semiconductor device modeling because of much lower parasitic resistances, capacitances, and inductances compared to measurement on a packaged device. It is usually done in a special chamber on top of a probe station since the chamber provides an enclosed environment to shield the device-under-test (DUT) from environmental lights and noises. There are mechanical manipulators outside the chamber to manually move the chuck under the wafer in *xyz* directions to accurately position probes on a test device. It is also possible to automatically measure many devices in a single die or in different dies on a wafer using computer-controlled probe station, which controls the chuck movement through servomotors. The chuck temperature, which offers an ambient temperature to the wafer, is set by a controller allowing measurement from –40°C to 200°C. The complete flow from on-wafer measurement to model parameter extraction is illustrated in Fig. 12.3.

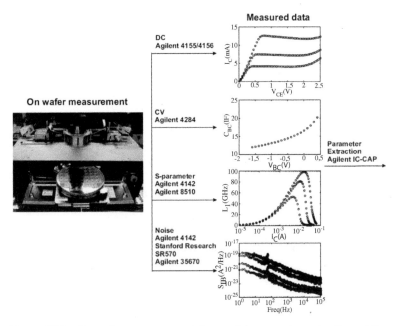

Figure 12.3 On-wafer measurement for model parameter extraction. After Ref. [1].

12.3.1 ICCAP Device Modeling: 1/*f* Noise Measurement Configuration

Low-frequency and high-frequency noise model equations are physics based and describe charge storage, including collector current spreading, up to very high current densities and in saturation bias dependence of internal base resistance, including both conductivity modulation and emitter current crowding temperature dependence of all elements. ICCAP is a dedicated toolkit for parameter extraction. The extraction routines are based on the device equations of model chosen to ensure that the extracted model parameter set gives a good physical representation of the device characteristics. Here, we describe the methods for accurate RF on-wafer measurement that are appropriate for production radio-frequency integrated circuit (RFIC) testing environments. The application of on-wafer probing to RFIC development, as well as to the evaluation of new semiconductor process technologies, is described. For RFIC measurements one needs to consider the following:

- Available probing technologies
- Probe selection
- Contacting techniques
- Testing of set hardware requirements
- Testing of structure layout
- System calibration
- De-embedding methods

Special emphasis is given to Si-based RFIC processes with attention to test structure layout criteria, substrate parasitic, and extraction of substrate effects, as well as optimum design of high-Q inductors. The testing approaches focus on evaluation of small-signal S-parameters noise parameters and load-pull large-signal test for nonlinear distortion. The material is presented to serve as a pragmatic guide to the techniques and requirements for accurate on-wafer test, the common pitfalls and how to avoid them, and the basic tools for performance and parameter extraction of both active and passive semiconductor devices in RF environments. A detailed technical description explaining the basics of modeling and parameter extraction of bipolar devices using the HICUM model has been developed [1].

Flicker noise or $1/f$ noise is an important noise source generated at low frequencies. The noise can cause severe effects on many RF applications. The challenge facing modeling engineers is the ability to measure the noise with precision. The Agilent ICCAP Device Modeling Configurations can be extended to add $1/f$ noise measurement capability with good accuracy. Controlled by ICCAP device modeling software, accurate and reliable measurements can be extracted and analyzed with the ICCAP Modeling Suite software. Figure 12.4 shows a block diagram of the $1/f$ noise measurement solution. The measurement configuration consists of the following major components:

- Agilent 4156C semiconductor parameter analyzer with two source monitor units (SMUs)
- Agilent 35670A dynamic signal analyzer
- Stanford Research SR570 low-noise current amplifier
- Digital voltage multimeter DMM
- Low-pass band filter
- G-S-G probes for on-wafer measurement (available from third-party companies such as Cascade Microtech)
- Cables and adapters

The following ICCAP software modules are basic requirements for a $1/f$ noise measurement solution:

- * 85199A IC-CAP software environment
- * 85199B IC-CAP analysis module
- * 85199D DC measurement drivers
- * 85199G noise measurement drivers
- * 85195B $1/f$ noise modeling package

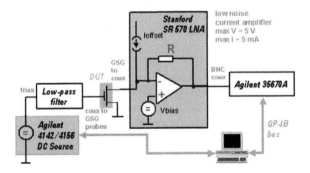

Figure 12.4   Block diagram of the $1/f$ noise measurement system.

12.4 Parameter Extraction Methodology

The first step is to specify the target parameters and the design variables, which, however, will depend on the particular technology to be developed. A list of major targets and variables could be as follows:

- Targets: threshold voltage, saturation current, off-state leakage/punch-through current, subthreshold swing, transconductance, gate/junction capacitance, and device lifetime, as well as derived circuit parameters such as gate propagation delay, static power, and noise margin
- Variables: channel length, gate oxide thickness, junction depth, channel doping profile, source/drain doping profile, and supply voltage, as well as all the major processing variables (diffusion/oxidation time and temperature, implant dose and energy, etc.) which influence the layer structures and profiles

For efficient circuit design, accurate transistor models are necessary. Historically the Ebers–Moll model is the first compact bipolar transistor model involving two back-to-back diodes. The evolution continued with the development of the GP, and more advanced models like VBIC, HICUM, and MEXTRAM. For accurate BJT modeling different extraction schemes have been proposed. Historically, the forward Early voltage has been extracted from the intercept with the V_{CE} axis of the extrapolated plot in the output characteristics for constant base voltage, as shown in Fig. 12.5. In the following, a method is given of how the model parameters are extracted (obtained from slopes/intercepts of straight lines) and optimized (obtained from least squares fits of data to the full model using the extracted values as initial values). The optimization is presented step by step on data sets which are limited to regions where the considered parameters are prevalent.

The reverse Early voltage has similarly been extracted as the intercept with the V_{CE} axis of the extrapolated output plot for constant base voltage. When the Early effect parameters are known, the current parameters ISS, IKF, IKR, βF, and βR are extracted. Typically, forward and reverse Gummel characteristics are used as shown in Fig. 12.6. The marked region is used for extraction of the saturation current IS from measurement data. Measurement is performed at zero base–collector voltage. The forward knee current IKF is extracted from the β ($=I_c/I_b$) versus $\log(I_c)$ plot as the value of I_c when β has dropped to half its peak value (Fig. 12.7).

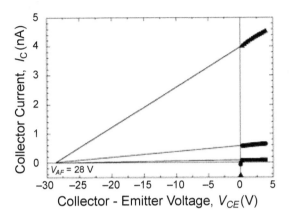

Figure 12.5 Traditional extraction of Early voltage.

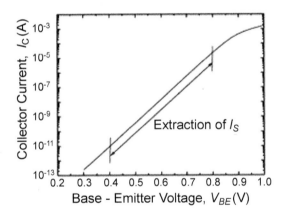

Figure 12.6 Extraction of the saturation current.

As an example, for the SPICE parameter extraction of a SiGe HBT, using HP-4145B semiconductor parameter analyzer one can obtain the device characteristics. From the experimental DC characteristics, respective model parameters can be extracted and compared. Figures 12.8 and 12.9 (forward Gummel and forward current gain) clearly show the drawbacks of the GP model to fit the experimental data, whereas the VBIC and HICUM models reliably reproduce the same. In Figs. 12.10 and 12.11 (reverse Gummel and the reverse current gain), a better fit of the VBIC than the GP model is observed. Figures 12.12 and 12.13 explicitly show that GP model is too weak to model quasi-saturation, avalanche, and self-heating effects, whereas VBIC and HICUM are capable of modeling these effects. A closer look reveals that in the forward high-current region HICUM model performs better, while the VBIC model is more suitable for both the forward and reverse regions of SiGe HBTs. A good fit on both, the DC and AC characteristics underline the validity of the extraction procedure.

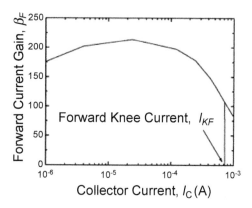

Figure 12.7 Extraction of the forward knee current.

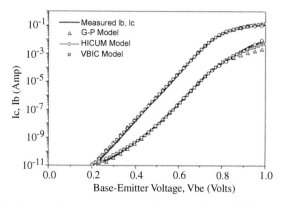

Figure 12.8 Forward Gummel plot of a SiGe HBT. After Ref. [2].

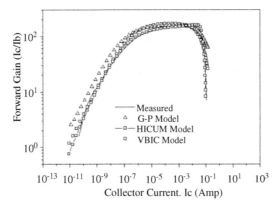

Figure 12.9 Forward DC current gain of a SiGe HBT. After Ref. [2].

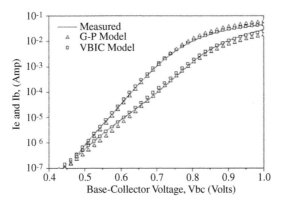

Figure 12.10 Reverse Gummel plot of a SiGe HBT. After Ref. [2].

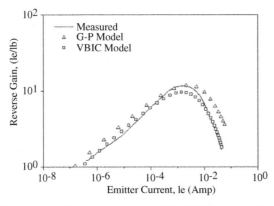

Figure 12.11 Reverse DC current gain of a SiGe HBT. After Ref. [2].

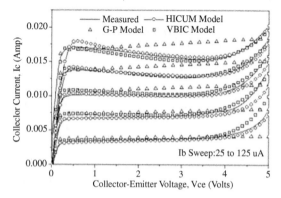

Figure 12.12   Forward output characteristics of a SiGe HBT. After Ref. [2].

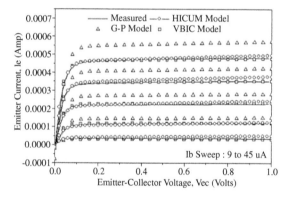

Figure 12.13   Reverse output characteristics of a SiGe HBT. After Ref. [2].

12.5   UTMOST

This following example adopted from Silvaco demonstrates the extraction of BSIM3 SPICE model parameters from ATHENA (process) to ATLAS (device) simulations for an n-type metal–oxide–semiconductor field-effect transistor (n-MOSFET). Using ATHENA, a metal–oxide–semiconductor (MOS) device in a salicide process is created and the device structure is interfaced to ATLAS with assigned variables to parameterize the structure. ATLAS is used to simulate I_d–V_g–V_{bs} and I_d–V_d–V_{gs} characteristics. Then UTMOST III is interfaced for the extraction of BSIM3 parameter set. In the process simulation a typical NMOS process flow is used. However, as the salicide process (titanium silicide) involves self-aligned silicide on the source, drain, and gate, the ATHENA/SILICIDES module is used. For the SPICE model extraction it is necessary to extract the gate oxide thickness in meter for use in UTMOST. Extract statements are used in process simulation to obtain these types of process parameters.

UTMOST from Silvaco includes a dedicated built-in BSIM3 parameter extraction algorithm, which is based on the extraction methodology proposed by Berkeley. The UTMOST BSIM3 extraction routine can perform both the measurements and the extraction sequence necessary for the extraction of an accurate scalable BSIM3 model. UTMOST supports the acquisition of semiconductor device electrical data from a wide range of measurement equipment, and the generation of model parameter sets from measured or simulated data, via parameter extraction and/or optimization methodologies. UTMOST extracts DC, AC, and transient device parameters for a very wide range of commercial and proprietary models for MOS, BJT, and other devices. The capability of UTMOST to use data from physically based simulators such as ATHENA and ATLAS is available as a component of Silvaco's Virtual Wafer Fab. Next, we describe the BSIM3 measurements using UTMOST and extraction routine. Using this routine UTMOST will be used to extract n-channel scalable models to devices with effective channel length of 0.6 μm.

Two ATLAS runs are used to parameterize using the set statement. The first ATLAS run simulates I_d-V_g characteristics at three substrate biases. The sequence of solve statements is to first save three solutions at each back bias with $V_d = 0.1$ V and $V_g = 0$ V. The second ATLAS run simulates I_d-V_d characteristics at three different gate voltages. A similar technique to the first run is used. Three solution files are saved at each gate bias with $V_d = 0$ V. These are then loaded in turn and the drain voltage ramped to "vdsmax".

The final stage of the example is to run UTMOST to extract the SPICE model. Important information about the structure such as gate oxide thickness and gate length is transferred to UTMOST using the results of extract statements in the ATHENA simulation. The log files from ATLAS are loaded and appended. UTMOST then fits the SPICE model to the complete I_d-V_g-V_{bs} and I_d-V_d-V_{gs} data sets. All the UTMOST parameters are stored in a file and then extract is used to print out the parameters of interest.

12.5.1 BSIM3

The BSIM3 extraction algorithm requires measurements from:

- A device with a large drawn length and a large drawn width (a Large device)
- Devices with a large drawn length and different drawn widths (W-array devices)
- Devices with a large drawn width and a range of drawn lengths (L-array devices)

The number of devices in the W-array or L-array is user-definable and UTMOST will accept a minimum or one device in each array. For a truly accurate scalable model it is recommended that a minimum of two devices exist in the L-array. A total of four sets of I–V measurements are necessary for the BSIM3 model parameter extractions. These measurement sets are detailed below:

Set 1: I_d–V_g measurements where the gate voltage is swept between defined minimum and maximum values. The source voltage is grounded and the drain voltage is set to a low value (i.e., 0.1 V). The bulk voltage is stepped, in six steps, between 0 V and a defined maximum.

Set 2: I_d–V_d measurements where the drain voltage is swept between defined minimum and maximum values. The source voltage is grounded and the bulk voltage is set to a low value (i.e., 0 V). The gate voltage is stepped, in five steps, between defined minimum and maximum values.

The minimum gate voltage is determined using the threshold voltage calculated from the data measured in measurement set 1.

Set 3: I_d–V_g measurements where the gate voltage is swept between defined minimum and maximum values. The source voltage is grounded and the drain voltage is set to a high value (i.e., 5.0 V). The bulk voltage is stepped, in six steps, between 0 V and a defined maximum.

Set 4: I_d–V_d measurements where the drain voltage is swept between defined minimum and maximum values. The source voltage is grounded and the bulk voltage is set to a high value (i.e., −3.0 V). The gate voltage is stepped, in five steps, between defined minimum and maximum values. The minimum gate voltage is determined using the threshold voltage calculated from the data measured in measurement set 1.

12.5.2 Parameter Extraction

Prior to any parameter extraction the user should specify values for some of the so-called BSIM3 elementary parameters like TOX, XJ, NPEAK, and NSUB. BSIM3 expert parameters can also be specified. The UTMOST BSIM3 extraction algorithm will now be described.

Step 1: Use the Large device linear region data from data set 1, and threshold voltages calculated from this data, to extract VTH0, K1, K2, U0 (optional), UA, UB, and UC.

Step 2: Use the W-array linear region data from data set 1, and threshold voltages calculated from this data, to extract K3, DW (optional), and W0.

Step 3: Use the L-array linear region data from data set 1, and threshold voltages calculated from this data, to extract NLX, DL (optional), DVT0, DVT1, and DVT2.

Step 4: Use the Large and L-array subthreshold region data from data set 1, and subthreshold slopes calculated from this data, to extract VOFF, NFACTOR, and CDSC.

Step 5: Use the L-array and W-array linear region data from data set 1 to extract RDS0 (optional) and RDSW.

Step 6: Use the Large and L-array data from data set 2 to extract VSAT, A0, A1 (if BULKMOD = 2), and A2 (if BULKMOD = 2).

Step 7: Use the Large and L-array data from data set 4 to extract KETA.

Step 8: Use the L-array saturation region data from data set 2, and the output resistances extracted from this data, to extract PCLM, DROUT, PDIBL1, PDIBL2, PSCBE1, and PSCBE2.

Step 9: Use the L-array subthreshold region data from data set 3 to extract ETA0, ETAB, and DSUB.

Step 10: After the extraction algorithm is completed UTMOST will optimize any selected BSIM3 parameters to the saturation region output conductance data for the L-array devices.

Scalable BSIM3 model parameters can easily be extracted with UTMOST built-in routine. Extraction examples for n-channel and p-channel devices from a 0.6 μm metal–oxide–semiconductor (CMOS) process were used to demonstrate the effectiveness of the extraction algorithm. The results presented here show how worst-case SPICE model parameters can be derived by the simulation of variations in the CMOS manufacturing process. This whole process can be automated as an experiment in Virtual Wafer Fab. Screenshots (Figs. 12.14 through 12.22) show the steps to be followed for SPICE parameter extraction using UTMOST.

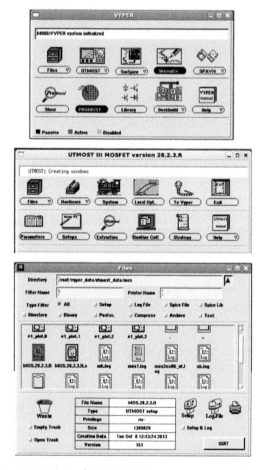

Figure 12.14 Screenshot of BSIM3 SPICE parameter extraction steps using UTMOST.

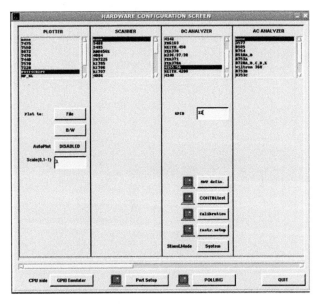

Figure 12.15 Screenshot of BSIM3 SPICE parameter extraction steps using UTMOST.

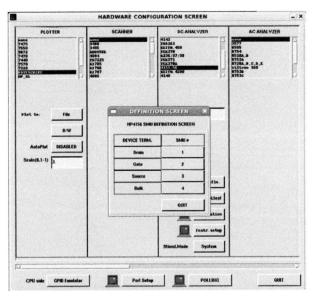

Figure 12.16 Screenshot of BSIM3 SPICE parameter extraction steps using UTMOST.

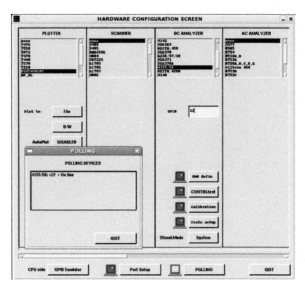

Figure 12.17 Screenshot of BSIM3 SPICE parameter extraction steps using UTMOST.

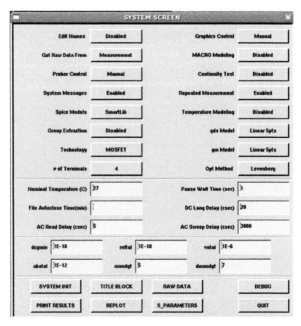

Figure 12.18 Screenshot of BSIM3 SPICE parameter extraction steps using UTMOST.

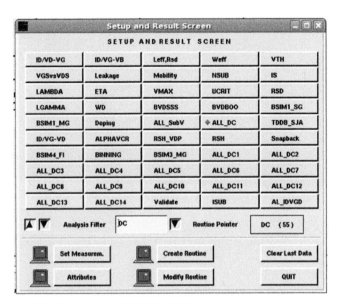

Figure 12.19 Screenshot of BSIM3 SPICE parameter extraction steps using UTMOST.

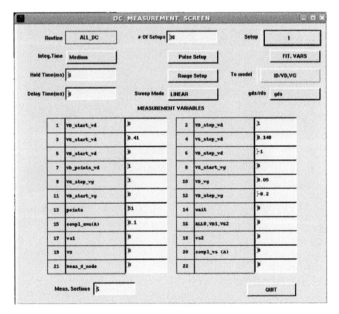

Figure 12.20 Screenshot of BSIM3 SPICE parameter extraction steps using UTMOST.

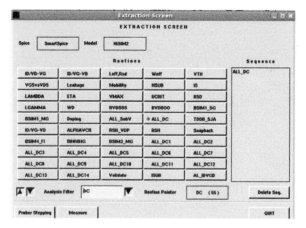

Figure 12.21 Screenshot of BSIM3 SPICE parameter extraction steps using UTMOST.

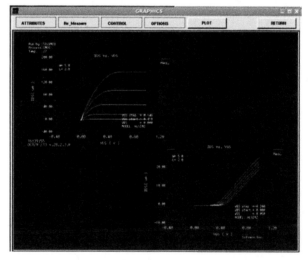

Figure 12.22 Screenshot of BSIM3 SPICE parameter extraction steps using UTMOST.

12.6 Summary

A compact model is needed substantially earlier in the time frame than traditionally available, depending on calibrated physical simulations for circuit designers. From an accurate process and device simulation early in the development cycle, the TCAD engineer can create the electrical device characteristics to derive a predictive compact model. A methodology for capturing process variability in SPICE models has been discussed. This chapter briefly reviewed available semiconductor device analyzer equipment, parameter extraction software, and methodologies used today for SPICE parameter extraction. Application of UTMOST for BSIM3 measurements and SPICE parameter extraction for MOSFETs has been described.

References

1. Wu, H.-C. (2007). *A Scalable Mextram Model for Advanced Bipolar Circuit Design*, PhD thesis, Technische Universiteit, Delft.

2. Chakravorty, A. (2005). *Compact Modelling of Si-Heterostructure Bipolar Transistors and Active Inductor Design for RF Applications*, PhD thesis, IIT Kharagpur.

Chapter 13

Technology CAD for DFM

At present, the design houses and wafer foundry are relatively independent of each other, but there is an increasing need for coupling and information exchange between the two. Most existing ECAD tools do not take into consideration the effects of the process variation. Presently, circuit designers perform optimization for catastrophic yield at a postsynthesis stage. Technology developers and circuit designers are generally separate entities and are very loosely linked by a set of physical device layout files and SPICE model parameters. This worked well before entering into the nanometer era because of the fact that transistor characteristics could be modeled unambiguously and statistical variations due to process fluctuations only represented a relatively small percentage of the nominal characteristics being modeled.

With 45 nm processes, it is imperative to develop a systematic TCAD-based methodology to design, characterize, and optimize manufacturability to increase yield. As the manufacturability of a process technology may be evaluated by the process window, defined as the area between the lower and upper limits of the critical process variables that yields acceptable device performance. The transition to 90 nm technology proved to be extremely challenging, and it is expected to worsen at 65 nm and 45 nm. Beyond pure CMOS, bipolar-based technologies target ever-increasing analog and mixed-signal IC performance, with a strong impact on parametric yield. This trend has spurred the need for innovative advanced process control methodologies reliant on an improved understanding of

Introducing Technology Computer-Aided Design (TCAD): Fundamentals, Simulations, and Applications
C. K. Maiti
Copyright © 2017 Pan Stanford Publishing Pte. Ltd.
ISBN 978-981-4745-51-2 (Hardcover), 978-1-315-36450-6 (eBook)
www.panstanford.com

the correlation between process variables and electrical device parameters.

Microelectronics fabrication is facing serious challenges due to the introduction of new materials in manufacturing and fundamental limitations of nanoscale devices that result in increasing unpredictability in the characteristics of the devices. The downscaling of CMOS technologies has brought about the increased variability of key parameters affecting the performance of integrated circuits. In silicon-based microelectronics, technology computer-aided design is well established not only in the design phase but also in the manufacturing process. Device design procedures are now more challenging due to high performance specifications, fast design cycles, and high yield requirements. Design for manufacturability and statistical design techniques are being employed to meet the challenges and difficulties of manufacturing of nanoscale integrated circuits in CMOS technologies. The methodology used to implement the information transfer from the analysis of process variations to the stage of circuit design is discussed in this chapter. In this chapter, the issues originating from different sources of process variability are discussed and modeling approaches which help to analyze the problems are presented. A 2D device simulation model is used to demonstrate the potential of the methodology. The process-aware SPICE model generated from simulations is presented.

13.1 Process-Aware Design for Manufacturing

The Synopsys Process-Aware Design for Manufacturing (PA-DFM) product family addresses the two major sources of variability in a design—one arising from stress proximity effects and the second from the spread of manufacturing process parameters across different die—and utilizes accurate physical models of the manufacturing process. The PA-DFM product family core products, Seismos, Paramos, and Fammos, enable designers to account correctly for manufacturing variability without major changes to the current design verification flow.

13.1.1 Seismos

As feature sizes continue to shrink beyond the 22 nm node, variability arising from advanced silicon technologies, such as strain engineering, increasingly affects circuit performance. Stress engineering has now become a critical factor in chip designs. As stress/strain engineering becomes pervasive for performance enhancement of advanced devices, study of the layout dependence of device characteristics has become essential. A process-aware design-for-manufacturing (DFM) solution is necessary for analyzing layout-dependent physical effects in advanced technologies. Seismos from Synopsys is a transistor-level design tool for the analysis of stress proximity effects in nanometer technologies.

13.1.2 Paramos

Paramos, a process-dependent SPICE Model extraction tool, is specifically designed to extract process-dependent SPICE model parameters for detailed analysis of circuits with process variations. The graphical user interface (GUI) allows users to develop an extraction strategy, run the extraction, and load the SPICE model card data into PCM Studio for visualization of extraction results.

Features:

- Provides process-related SPICE parameters for detailed analysis of circuits with process variations, thereby closing the design for manufacturing (DFM) gap
- Creates self-consistent process-dependent compact SPICE models with the actual process parameter variations as explicit variables
- Enables designers to comprehend the impact of manufacturing issues on design
- Allows designers to simulate the impact of process variability (statistical or systematic) on circuit performance for design margin improvement or parametric yield sensitivity analysis
- Allows process engineers to perform design-specific process centering

13.1.3 Fammos

The third product in the PA-DFM family is Fammos, the first special-purpose technology computer-aided design (TCAD) PA-DFM tool that analyzes stress evolution for the entire fabrication process of interconnects. Fammos performs 3D backend process simulations using design database and process recipes. With specialized algorithms for fast 3D structure construction, mesh generation, and equation solving, Fammos predicts interconnect stress distributions from multiple stress sources and accounts for proximity effects. It employs a set of physics-based models to evaluate reliability failures. Using the Sentaurus Workbench user interface, Fammos facilitates technology explorations with parameterized input files and scheduled run splits.

Features:

- Generates 3D structures using design database and process steps
- Analyzes stress evolution for the entire fabrication process
- Accounts for multiple stress sources and proximity effects
- Performs technology explorations for yield and reliability improvements

13.2 TCAD for Manufacturing

TCAD tools provide detailed physical insight into achieving the optimum process performance affecting yield. TCAD represents our physical understanding of processes and devices in terms of computer models of semiconductor physics. The scope of conventional TCAD generally includes front-end process modeling and simulation, implant, diffusion, oxidation, numerical device modeling and simulation for electrical characteristic simulation, topography modeling and simulation, deposition, lithography, and etching. Over the last decade, there has been steady progress in the development and acceptance of CAD programs for technology development—integrated tools for process analysis and device/circuit design. In extending TCAD into the manufacturing and design area, one needs to consider the effects of process variability in the circuit design phase. DFM involves yield analysis and statistical process control.

The Sentaurus TFM suite, which includes PCM Studio and PCM Library, provides a powerful environment for capturing multivariate process–device–circuit relationships in process compact models (PCMs), allowing a fast turnaround for identifying and analyzing factors that cause parametric yield loss in manufacturing. Derived from systematic TCAD simulations, PCMs encapsulate relationships between process variations and device circuit performance through a set of analytic functions.

Process-aware design provides a direct link from process variation to circuit response and is a methodology that assigns variability of individual process and its effects to circuit performance. Two main tasks process-aware design needs to perform are (i) to quantify the parametric variability and link it to a specific origin and (ii) to identify a suitable compact model (CM) to link process variation to circuit performance. Currently, fabless companies who have limited or no control over the process use the design kits. Fabless companies are considering the opportunity of changing their business model as the process research and development are facing new challenges. Process-aware circuit simulation becomes more evident for the fabless companies who are isolated from the manufacturing process. Process-aware circuit simulation may provide an invaluable tool for fabless companies to deliver robust designs.

Process variability has become the primary concern with regard to manufacturability and yield. With scaled feature dimensions in the nanometer range, parametric variations become a dominant yield loss component. Reasons for yield loss include random defects due to a dominant yield loss mechanism, defects due to inadequate or incorrect lithography, and defects due to pattern (or design). As device dimensions shrink, the sensitivity of device performance to process variation also increases. For complex nanometer designs, yield might depend more on design attributes than on total chip area.

Although the problem of variability is becoming more pronounced in advanced technologies, it has always been present and is being addressed by the industry following different methodologies. As process variations are present, there is usually a distribution of values for device characteristics such as saturation current, leakage current and threshold voltage. There are different ways to extract corner, models and the standard approach is based on electrical measurements. However, the standard approach needs a frozen process and a large set of measurement data. It is thus not applicable in an early stage of process development when stable data are not yet available.

The problem in extracting SPICE parameters is the definition of the typical (or nominal) device. The main procedure for years has been to choose a "golden wafer" from which the parameters are extracted. The golden wafer provides all the electrical and process in-line measurements close to the nominal value. The variability of the process is then accounted for with the generation of corner models which represent worst-case models for the technology in which a designer should test the circuits for functionality. The use of corner models for worst-case testing of a circuit in the design stage has several limitations. The main drawback of using corner models is that it is not possible to link them to a process step. Corner models are essentially being a pass/no pass test and do not provide any information to the designer on what to look at in the case of circuit failure.

The scaling of technology has reached a point at which the extremely small dimensions of the devices are becoming a challenge for manufacturing. The manufacturing tolerances are becoming a challenge for devices and circuits, threatening time to market and yield. Information transfer between process and design engineers is required for robust designs. With feature sizes approaching the nanoscale, unavoidable fluctuations that are intrinsic to the manufacturing process are becoming important. Polysilicon/metal edge grains, gate oxide thickness and permittivity nonuniformities, and photoresist edge roughness are among the major sources of fluctuations.

Systematic variability arises from process–design interactions, proximity effects in lithography, chemical-mechanical polishing, and stress-related effects that are position dependent. This kind of variability affects transistors in the same die in different ways, but the die-to-die impact is the same. Parametric variability includes the drift of nominal process conditions such as implantation energy, dose and temperature. It affects die-to-die, wafer-to-wafer, and lot-to-lot device characteristics. Parametric variability causes the mean value of transistor parameters to change during time or among different chips or different lots.

Random variability encompasses all kinds of variation that are not properly understood or measured. In particular, it includes random fluctuations that are beyond process control possibilities. Line edge roughness and random dopant fluctuations belong to this category. Moreover, this kind of variability is statistical in nature and results in average transistor parameters such as threshold voltage, saturation, and leakage currents. Process variation sources include:

- Wafer: Topography, reflectivity
- Reticle: CD error, proximity effects, defects
- Stepper: Lens heating, focus, dose, lens aberrations
- Etch: Power, pressure, flow rate
- Resist: Thickness, refractive index
- Develop: Time, temperature, rinse
- Environment: Humidity, pressure

The traditional approach to manage fluctuations in manufacturing has been to limit the variability of individual process step. At the design level, corner models are typically used to take process variability into account. These models represent the worst cases for fabricated devices and are used by designers to test the products for functionality in extreme cases, leading to only a go/no-go test. Predictive potentials of TCAD depend on process variations which get increasingly critical with device downscaling into the nanometer range. For example, phenomena such as line edge roughness and random dopant fluctuations broaden the device parameter distributions, thus requiring statistical analysis.

DFM is a technique for addressing the producibility issues early in the design cycle, and integrating manufacturing concerns and considerations into a design to obtain a more producible product. Design for manufacturability consists of three stages: physical design, resolution enhancement techniques, and design-driven techniques. IC design methodologies typically target nominal designs. However, defects and variations in the IC manufacturing process can make a circuit behave substantially different from the nominal design. Yield is the percentage of manufactured products that meet all performance and functionality specifications. Parametric yield loss usually refers to the effects on circuit performance caused by process variations.

DFM techniques counter measures against manufacturing difficulties. Different DFM tools address the problems derived from variability in the manufacturing process. The most common ones address systematic variability caused by manufacturing limitations. DFM consists of three categories: physical design, resolution enhancement, and design-driven techniques. The first two are concerned with how to control and reduce the effect of process, voltage, and temperature variations. Design-level considerations for parametric yield, a dominant factor to determine the total manufacturing yield, are critical and the statistical design methodology needs to be reinforced. Analysis of process variability in circuit simulation enables designers to enhance the design robustness. The effects that a specific process variation has on device variability and on the final circuit performance are the key to improve yield. In design, the traditional approach to ensure design robustness against variability is represented by corner models that lump together all sources of variability.

DFM and design for yield require electronic design assistant (EDA) tools that fully comprehend the impact of novel technology concepts and their influence on the process variability of devices and interconnects. A collaborative platform for DFM aims to meet this challenge by joining process and circuit simulators as well as a set of process characterization experiments that are needed to enable quantitative DFM. At the 22 nm node and beyond, process variability will increase with feature scaling and the introduction of new materials and techniques such as strain engineering. With processing costs increasing dramatically for advanced technologies, the economic necessity for accurate predictive TCAD is apparent.

13.3 TCAD for DFM

State-of-the-art TCAD tools, which not only provide deeper insight into novel devices but also have a high-level predictive power, may be used successfully to characterize process variability through simulations. TCAD is inexpensive and faster than an experimental analysis that requires costly design of experiments (DOE). Moreover, using simulation, it is possible to analyze quantities not otherwise accessible by measurements. The evaluation of doping profiles in very small structures as well as electric fields in an operating device is only possible through simulation. In the process-aware design exercise using TCAD, device characteristics generated using simulation is substituted for the measured data. SPICE models are extracted in the usual way from simulated devices to perform circuit simulation. In TCAD, it is possible to precisely control parameters during simulation that are difficult to access through metrology. However, the accuracy TCAD simulation depends on the calibration of various models.

Figure 13.1 shows a typical simulation flow. The starting point is a technology process flow which creates the device geometry through process simulation and provides the topography and doping profiles of the final device. The device under study is remeshed in order to refine the areas of interest and eventually the device simulation run is performed to provide the device electrical characteristics. The simulation-based methodology starts from the analysis of process variability to the extraction of a process-aware SPICE (PA-SPICE) CMs that may be used in circuit simulation.

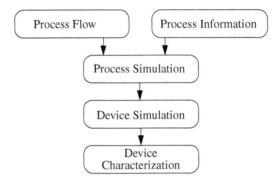

Figure 13.1 Typical TCAD simulation flow from process information to device characteristics.

Figure 13.2 shows the simulation flow from process description to final circuit simulation. Two different levels of modeling can be distinguished: Process optimization links the process variability to the device characteristics, while process-aware design brings the information on process variability to the circuit simulation stage to optimize the design. Next, the methodology to enable a process-aware circuit simulation is applied to a generic complementary metal–oxide–semiconductor (CMOS) process. The Sentaurus TCAD simulation suite may be used to perform all the relevant simulations. In Fig. 13.2 the simulation flow is shown. Two different levels of modeling are involved: (i) The process optimization links the process variability to the device characteristics, and (ii) process-aware design brings the information on process variability to the circuit simulation stage to optimize the design. As the entire space of process variables is very wide in a manufacturing process, covering entire process variable set is extremely difficult even in simulations.

It is therefore necessary to limit the optimization study using the DOE to those process parameters that have a major impact on the final device, reducing the dimensionality of the input space.

The main variables that affect the manufacturing process output are found by a screening experiment on the process input variables. Screening experiments are inexpensive design used in the exploratory phase of a process development to gather information on the process step. A sensitivity analysis on the parameter of interest is considered. In the deterministic context, sensitivity analysis is performed by varying one parameter at a time in a small interval around its nominal value, while all others are kept fixed in order to determine its impact on the process output. Once the main factors affecting a process have been determined, it becomes important to study their interaction on the final device characteristics and eventually on the circuit.

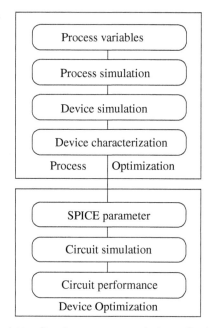

Figure 13.2 Simulation flow from process to device to final circuit simulation.

A factorial design is planned to achieve this from different available possibilities. A full factorial design approach requires the highest number of experiments as it requires different experiment for all combinations. However, if the full factorial approach becomes impractical, fractional factorial designs which are subsets of the full factorial can be used. The information gathered through factorial experiments on the relevant process parameters can be used to find the functional relationships between input parameters and output characteristics of the process under analysis. The functional relationship is then used for process optimization. Response surface models (RSMs) are classes of models that are well suited for process optimization and are also called PCMs.

Once a simulation-based methodology to identify the main process parameters affecting device characteristics is developed, it is important to bring the information of the process variability to the circuit design stage. It is now possible to quantify the impact of a process step variation not only on the threshold voltage of a single transistor but on the response of the circuit. The use of TCAD to extract PCMs to account for variability is demonstrated. Process parameters are global parameters that affect the full circuit response. It is possible to identify two approaches to process-aware analysis of circuit performance. The first approach applies response surface modeling directly in the SPICE parameter space. The idea is to extract a SPICE model whose parameters are not fixed values, but functions of process input factors.

The extracted PA-SPICE model is compatible with other SPICE simulator tools, which allows the designer to use the process parameters in the same way as normal library parameters [1]. The main advantage of this methodology is that the PA-SPICE model requires a single extraction step. It can be used on different circuits to evaluate the sensitivity to process variations. PCMs may also be used as a figure of merit of a particular circuit. For example, starting from the DOE with n experiments, n SPICE models are extracted each reflecting a different process condition. Then the circuit is simulated n times using different model cards and the PCM linking the process parameters to the circuit responses is built. The circuit response PCM can then be used for yield considerations and to analyze process windows that meet product specifications. However, the generated PCM is applicable only to the circuit from which it was generated.

13.4   Process Compact Models

At the core, traditional TCAD has only process and device simulations; however, the TCAD environment can be extended to form a self-contained system, from technology development to circuit performance extraction. The traditional role of TCAD has to be extended to the advanced high-level approach and fundamental low-level approach. The high-level approach is required for the tightly coupled design and technology development. The advance CM has to keep pace with the technology development. A CM provides a link for process-device-circuit interdependence from a lower-level model. In addition, design for reliability (DFR) and DFM must be included.

The methodology used to implement the information transfer from the analysis of process variations to the stage of circuit design is shown in Fig. 13.3. Sentaurus TCAD tools for process and device simulation have been used to design and optimize a typical 45 nm CMOS process with double-halo implantation and a combination of conventional rapid thermal annealing (RTA) with laser anneal for low leakage/threshold voltage and high-gain devices. The high dopant activation and low transient-enhanced diffusion (TED) aspects of this process are fully captured with Sentaurus Process simulation tools. Sentaurus Device is used to simulate DC and AC characteristics, and the Sentaurus Workbench tool Inspect is used to extract device parameters such as V_t, I_{on}, I_{off}, and radio-frequency (RF) parameters. Prior to systematic TCAD simulations, a sensitivity analysis is carried to determine the critical process variables and suitable ranges for the experimental design. A full-factorial DOE is executed and the PCM is extracted. From calibrated TCAD it is possible to generate an RSM, and find the optimized 45 nm CMOS process. The process is optimized (to select a stable process window) with respect to threshold voltage, drive current, and leakage.

Toward extended TCAD, in process modeling, generally a systematic DOE run is performed. A DOE can be systematically set up to study the control over process parameters and an arbitrary choice of device performance characteristics. The models developed from DOE are known as PCMs, which are analogous to CMs for semiconductor devices and circuits. PCMs may be used to capture the nonlinear behavior and multiparameter interactions of manufacturing processes. SPICE process compact models (SPCMs) can be considered as an extension of PCMs applied to SPICE parameters.

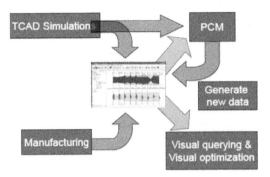

Figure 13.3 Flow diagram for PCM creation and evaluation.

By combining calibrated TCAD simulations with a global SPICE extraction strategy, it is possible to create self-consistent process-dependent compact SPICE models, with process parameter variations as explicit variables. This methodology brings manufacturing to design, so that measurable process variations can be fed into design as shown in Fig. 13.3. Next, we discuss a general formulation of the device optimization problem that is composed of the selected device design parameters and constraints. Thereafter the idea behind the yield maximization process is established, and the problem is formalized by the yield maximization technique. The full simulation steps followed are:

- Process simulation setup
- Electrical response evaluation (I_d–V_g, I_d–V_d, C–V, and RF)
- Process optimization using a PCM
- Sensitivity, DOE, and PCM validation
- Analysis using PCM
- PCM generation
- Manufacturability and yield analysis

The process-to-device-simulation setup steps are:

- Process simulation (using Ligament and Sentaurus Process)
- Mesh generation for the device simulator (using Sentaurus Structure Editor)
- Device simulation (with Sentaurus Device)

- Data analysis and postprocessing (using Inspect and Tecplot)
- State-of-the-art polynomial model for PCM creation

The model selection steps are:

- The three-stream pair diffusion model selected for the conventional RTA steps and the five-stream react diffusion model for the nonequilibrium of dopant–defect pairs during the fast ramp rates
- Electrostatic simulations using the hydrodynamic transport and density gradient models for electrons and holes
- Nonhomogeneous thermal distribution performed with the LASER subcommand within the Diffusion section

13.4.1 Process Parameterization

To demonstrate process optimization using PCM Studio, one device parameter, for example, threshold voltage, is chosen and the process is optimized with respect to V_t. As an example, we optimize the device performance by minimizing threshold voltage which mainly depends on the following parameters: halo implant dose (Halo_Dose) and extension implant dose (Ext_Dose), gate length (L_g), gate oxide thickness (G_{ox}), and peak temperature for RTA which modifies the doping concentration in the channel region, as shown in Fig. 13.4. The optimization problem consists of finding the best combination of the above parameters that produces the desired threshold voltage. The visual optimization procedure allows one to put constraints on the input parameters which, however, are motivated by the manufacturing considerations.

13.4.2 Process Calibration

Validating the impact of parameter variation on the device responses has an important role for the PCM generation. To have an accurate PCM it is important to have a numerically stable process flow. This is typically performed with local sensitivity analysis, such as one parameter is varied, while the others are kept constant. The local sensitivity analysis also serves for screening, that is, determining which parameters are most important. Sensitivity analysis determines the amount of change produced in the model responses by a local change in a model parameter, usually by computing the partial derivatives. Uncertainty analysis uses global sensitivity analysis to determine the variances of the responses caused by variance in the process parameters. Yield analysis performs an uncertainty analysis with constrained responses with estimation of "good"/total ratio.

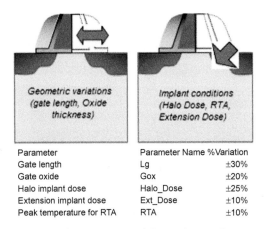

Parameter	Parameter Name	%Variation
Gate length	Lg	±30%
Gate oxide	Gox	±20%
Halo implant dose	Halo_Dose	±25%
Extension implant dose	Ext_Dose	±10%
Peak temperature for RTA	RTA	±10%

Figure 13.4   Process and geometry variability and range for optimization [1].

13.4.3   TCAD Validation

TCAD validation steps are:

- Smoothness analysis: For an accurate PCM a numerically stable process flow is important.
- Sensitivity analysis: Parameters are varied one at a time.
- Screening: Which parameters are most important is determined.
- Assessment of nonlinearity: Are polynomial expressions sufficient to capture the effects?

13.4.4   PCM Simulation

Many experiments are necessary for good PCM characterization. A particular choice of DOE is a trade-off between the value of knowledge and the cost of experimentation. Generally a practical DOE often targets main effects and their interactions, neglecting the influence of higher-order interactions. Figure 13.5 shows a normalized histogram plot summarizing the sensitivity analysis for the critical process steps. The variation of each output parameter for the specified input range is normalized to the maximum value; that is, the *y* axis range is 0 to 1. Normalized process variability is used to find out the most important and sensitive process, as shown in Fig. 13.5.

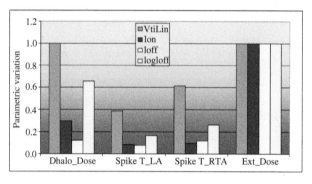

Figure 13.5 Sensitivity analysis of process variability for a 45 nm MOSFET [1].

A very useful way of visualizing and analyzing PCM-generated data is the parallel coordinate plot which plots multivariate data in a single representation. It is created by mapping coordinates in a multidimensional space onto a set of parallel axes, one for each input and output parameter. A line connects the corresponding coordinates. In particular, the parallel coordinate plot is an ideal environment in which to perform a visual optimization. Figure 13.6 shows the parallel coordinate plot. The process is optimized with respect to threshold voltage, saturation current, and I_{off}.

PCM is used to determine the sensitivity of the process parameters visually as shown in Fig. 13.7. The local sensitivity and global sensitivity can be determined by adjusting the sliders.

The combination of SProcess, SDevice, PCM Studio, and Sentaurus Workbench forms a powerful DFM TCAD environment. In this study, a total of 1200 experiments were generated. The process and device simulation results are subsequently used as the basis for generating a PCM, which encapsulates the relationships between input (design) and output parameters. The PCM automatically correlates design parameters to the tolerances. Parallel coordinate plots link the simulation results to the design variation. The parameter values and ranges indicate whether the domain has been covered sufficiently. Figure 13.8 is a parallel coordinate plot that links the simulation results to the design optimized with respect to threshold voltage, saturation current, and I_{off}.

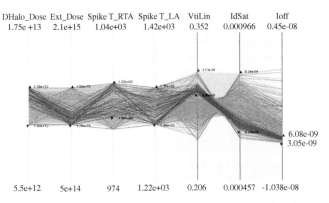

Figure 13.6 Parallel coordinate plot. The process is optimized with respect to threshold voltage, saturation current, and I_{off} [1].

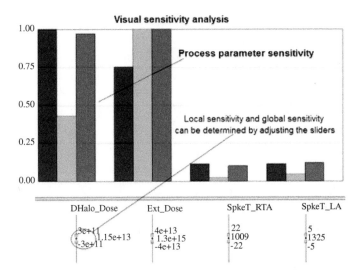

Figure 13.7 Visual sensitivity analysis. The local sensitivity and global sensitivity can be determined by adjusting the sliders [1].

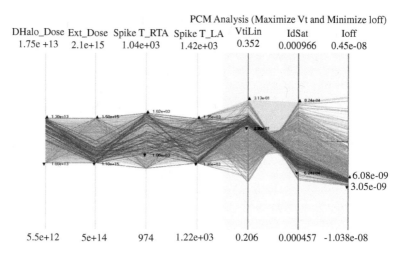

PCM Analysis (Maximize Vt and Minimize Ioff)

DHalo_Dose	Ext_Dose	Spike T_RTA	Spike T_LA	VtiLin	IdSat	Ioff
1.75e +13	2.1e+15	1.04e+03	1.42e+03	0.352	0.000966	0.45e-08

| 5.5e+12 | 5e+14 | 974 | 1.22e+03 | 0.206 | 0.000457 | -1.038e-08 |

6.08e-09
3.05e-09

Figure 13.8 Parallel coordinate plot that links simulation results to the design optimized with respect to threshold voltage, saturation current, and I_{off} [1].

It is observed that several process variants exists for the given specification limits of the device target. Thus it is important to find the most robust process condition for known process variations around each process condition using PCMs so that the stability of each process condition can be quantified. Determination of the most stable process condition is shown in Fig. 13.9. Marked process conditions indicate low sensitivity of the device characteristics to the variations in corresponding set of process. Similarly, when several process variants exists over a wide-design-limit target, it is necessary to sort the process conditions on the basis of their sensitivities, taking into account the process variations around each condition. Using PCMs, the sensitivity of each process condition can be quantified.

So far, we only optimized device performance. Let us now add aspects of manufacturability, that is, the minimization of the impact of parametric variations. We introduce the resulting variance in device characteristics as an optimization constraint. Figure 13.10 shows the device specification limits. Quantification (0–1) of yield is shown in Fig. 13.11. The maximum yield for a process condition is 1. The DFM/PCM simulation example demonstrates how to optimize a process and reduce the process development time by reducing the number of costly and time-consuming design iterations.

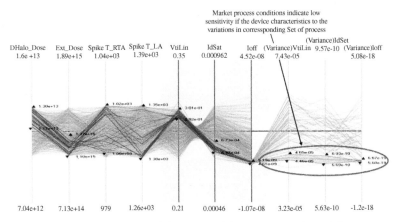

Figure 13.9   Determination of the most stable process condition [1].

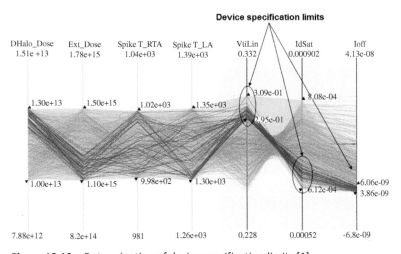

Figure 13.10   Determination of device specification limits [1].

13.5   Summary

The manufacturing challenges introduced by scaling of semiconductor technologies require a strengthening of the communication process between process engineers and designers. This chapter focused on problems caused by variability in the manufacturing process, and it presented modeling capabilities which help to analyze the problems and ultimately to counter them by delivering robust designs to the foundry. Challenges in DFM, how to incorporate process variations in the design, and how to identify and characterize the main sources of transistor performance variation were discussed.

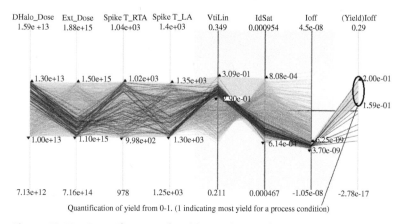

Quantification of yield from 0-1. (1 indicating most yield for a process condition)

Figure 13.11 Quantification of yield. 1 indicates the maximum yield for a process condition [1].

The importance of variability in semiconductor production process and the need of a process-aware design approach were outlined. TCAD was shown to be an essential source of reliable information to quantify the variability in a process compared to traditional experimental techniques. We discussed the methodology to study the parametric variability of a process and to bring the relevant information to the design space.

Reference

1. Maiti, T. K., private communication.

Chapter 14

VWF and Online Laboratory

An integrated circuit (IC) chip design starts with the product specification, followed by the front-end and backend designs. IC design can be broken down into two categories, digital and analog. Mixed-signal design works with both digital and analog signals simultaneously. Digital IC design has clearly defined steps and procedures to produce circuits. Analog IC design is performed at the circuit level and the designs are more complex in nature. An analog design engineer needs to possess a strong understanding of the principles, concepts, and techniques involved in the electrical, physical, and testing methodologies employed in circuit fabrication. In general, as the EDA vendors who provide electronic computer-aided design (ECAD) tools for the design houses and TCAD vendors who provide TCAD tools for the manufacturing are separate entities, the link between the design house and the manufacturing has so far been poor. For the design houses, still the current practice is to rely on the manufacturer to provide the model parameters necessary for design purposes.

The IC CAD framework typically consists of four levels of modeling, viz., technology level, device level, circuit level, and system level. Technology-level modeling is related to the device structure and doping profile and their dependence on process variations, as well as the resulting electrical characteristics. Device-level modeling is related to the description of transistor terminal characteristics

Introducing Technology Computer-Aided Design (TCAD): Fundamentals, Simulations, and Applications
C. K. Maiti
Copyright © 2017 Pan Stanford Publishing Pte. Ltd.
ISBN 978-981-4745-51-2 (Hardcover), 978-1-315-36450-6 (eBook)
www.panstanford.com

normally expressed in terms of compact model (SPICE) parameters. Circuit-level modeling refers to the solution of large linear/nonlinear systems of equations by various matrix solution techniques. System-level modeling refers to the analysis of the behavioral blocks that make up a given system. Traditionally, the above levels of abstraction are relatively independent and require different tools.

In the design phase, for a given technology node, ECAD tools are used to develop (synthesize) the logic design from a high-level hardware description language (HDL), and the circuit net list is extracted from the logic functional description, and then the layout is extracted from the circuit- and logic-level descriptions. Once a set of masks has been designed, combined with the process recipe, it goes for manufacturing in a fabrication facility. Technology characterization, device parameter extraction and testing are then performed on the fabricated devices/circuits. Conventional IC manufacturing thus follows the loop *design—manufacturing—characterization—verification*. This loop can be very time consuming and costly if a first-time silicon success is not achieved and more so if a new technology is needed to be developed using this iterative experimentation. It is obvious if virtual wafer fabrication (VWF) could be used to supplement the real wafer fabrication (RWF) experimentation. VWF can then bridge the gap between the design and manufacturing which is becoming increasingly vital in the deep submicron GSI era and also between the ECAD and TCAD tool users, in general. There is always a need for continuous feedback from analog designer that help the circuit designer in the early stages of the design. VWF provides TCAD services to device researchers, technology developers, and circuit designers.

Technology developers and circuit designers are generally separate entities and are very loosely linked by a set of physical device layout files and SPICE model parameters. This worked well before entering into the nanometer era, due to the fact that transistor characteristics could be modeled unambiguously and statistical variations due to process fluctuations only represented a relatively small percentage of the nominal characteristics being modeled.

In general, as the EDA vendors who provide ECAD tools for the design houses and TCAD vendors who provide TCAD tools for the manufacturing are separate entities, the link between the design house and the manufacturing has so far been poor. For the design houses, still the current practice is to rely on the manufacturer to provide the model parameters necessary for design purposes.

Current trends place demands for new design methodologies as well as new design tools. An advanced TCAD framework should have the following capabilities:

- Developing new process technology for device design and optimization
- Calibrating the process for a technology node so that the developed models can be used in the new process
- Providing process windows for given device performance targets through design of experiments (DOE) for the final optimal design
- Providing a first-order approximation on device performance in relation to process variables
- Extracting circuit performance from virtual wafer experiments

This means that the transistor structure and processing information need to be considered in the early stage of a design before it is fabricated. SPICE parameters can be extracted from the virtual device *I–V* characteristics. The SPICE parameters can be obtained through parameter extraction on the basis of the simulated electrical characteristics obtained through technology characterization. VWF is a collection of several software modules, each responsible for part of a device simulation. VWF is a feature-rich tool for performing DOE and optimization experiments seamlessly which integrates Silvaco simulators and postprocessing tools into one graphical user interface (GUI). The four principal components are as follows:

- **DeckBuild** provides a "home base" for the user to load examples or define code used to fabricate and simulate devices.
- **ATHENA** carries out simulation of the device fabrication. ATHENA performs various processing steps such as oxidation, material deposition, diffusion, and etching.
- **ATLAS** is responsible for device simulation (basically performs the device electrical characterization). The user specifies terminal voltages, for example, and ATLAS generates current–voltage characteristics, and displays equipotential lines, electric field lines, and charge carrier concentrations (among various other possibilities).

- **TonyPlot** performs the plotting of the results of the simulations.

Silvaco VWF interactive tools provide a versatile environment for using semiconductor TCAD tools (Fig. 14.1). VWF tasks are performed by physically based process and device simulators which are available in the interactive tool environment. VWF is comprised of three basic component sets such as, core tools, interactive tools and automation tools. Core tools simulate either a semiconductor device being processed or a semiconductor device being tested electrically. A DOE is performed to determine a response based on differing inputs. For example, the transistor electrical characteristics are closely coupled with the doping profiles and layer structures, and very often, trade-off must be taken for different design targets. For modeling a diffusion process, one may wish to vary the effects of implant dose, screening oxide thickness, drive-in temperature, and drive-in time to see the effects on the implanted dopant profile.

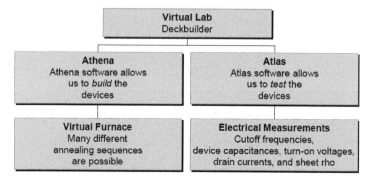

Figure 14.1 Flow diagram for Silvaco VWF.

The VWF interactive tools are designed to be used interactively in the construction of a single input deck. GUI based, they make the job of constructing an input deck more efficient. The interactive tools may be used either in conjunction with a set of files or as integral components of the surrounding VWF automation tools. The VWF automation tools enable a user to perform large scale experimental studies to create results for subsequent statistical analysis. Next, we present Silvaco Virtual Wafer Fabrication (VFW) as an example.

Silvaco VFW has two modes of operation, database mode and file mode. In order to demonstrate the DOE, one may import an existing Silvaco example deck that can be used. A simple diffusion example ATHENA diffusion called *andfex01.in* is used for the DOE, as shown in Fig. 14.2. One needs to make some of the constants defined in the deck into variables such as dose, energy, time, and temperature as variables, as shown in Fig. 14.3. The extracted variable in this case study is junction depth, "xj."

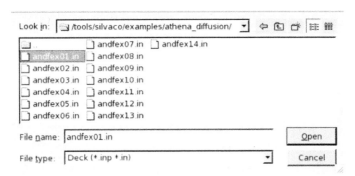

Figure 14.2 Silvaco VWF import deck selection window. After Ref. [1].

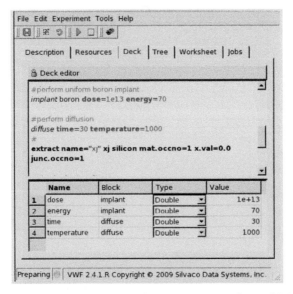

Figure 14.3 Silvaco VWF variables defined in the deck. After Ref. [1].

One can verify the status of design by clicking on the tab marked *tree*. Initially, it will only have five steps, as shown in Fig. 14.4. By selecting Edit and then Design from the top menu, one can enter the experiment design window. Figure 14.5 shows a filled-out example set of settings for the experiment being designed. The final tree view should now look like what is shown in Fig. 14.6.

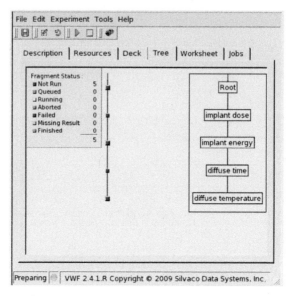

Figure 14.4   Silvaco VWF initial tree layout. After Ref. [1].

Design:	2 Level Full factorial			
	Variable	Initial	Low	High
1	☑ implant dose	1e+13	1e+13	2e+13
2	☑ implant energy	70	60	70
3	☑ diffuse time	30	20	30
4	☑ diffuse temperature	1000	1000	1100

OK	Cancel

Figure 14.5   Silvaco VWF experiment design window. After Ref. [1].

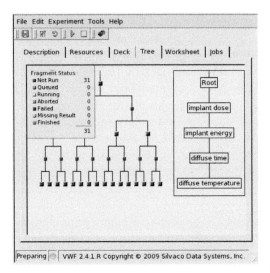

Figure 14.6 Silvaco VWF final tree layout. After Ref. [1].

Once the simulations have run, there is couple of ways to verify the output. One can select any of the bottom squares to look at a plot of the data. Otherwise, the easiest way is to export the data to a program called Spayn. Select tools and then Spayn needs to have the data exported to Spayn after clicking OK on the next dialog window. Figure 14.7 shows the data that was exported to Spayn. One can use the tools built into the program to produce other graphs and do statistical analysis. It is to export the data to a comma delimited file for easy import into Microsoft Excel.

File Edit View Tools Simulation Help

	dose	energy	time	temperature	xj
1	1.000e+13	7.000e+01	3.000e+01	1.000e+03	7.006e-01
2	1.000e+13	7.000e+01	3.000e+01	1.100e+03	1.164e+00
3	1.000e+13	7.000e+01	2.000e+01	1.000e+03	6.768e-01
4	1.000e+13	7.000e+01	2.000e+01	1.100e+03	1.022e+00
5	1.000e+13	6.000e+01	2.000e+01	1.000e+03	6.353e-01
6	1.000e+13	6.000e+01	2.000e+01	1.100e+03	9.886e-01
7	1.000e+13	6.000e+01	3.000e+01	1.000e+03	6.597e-01
8	1.000e+13	6.000e+01	3.000e+01	1.100e+03	1.134e+00
9	2.000e+13	6.000e+01	2.000e+01	1.000e+03	6.597e-01
10	2.000e+13	6.000e+01	2.000e+01	1.100e+03	1.031e+00
11	2.000e+13	6.000e+01	3.000e+01	1.000e+03	6.849e-01
12	2.000e+13	6.000e+01	3.000e+01	1.100e+03	1.184e+00
13	2.000e+13	7.000e+01	2.000e+01	1.000e+03	7.039e-01
14	2.000e+13	7.000e+01	2.000e+01	1.100e+03	1.062e+00
15	2.000e+13	7.000e+01	3.000e+01	1.000e+03	7.273e-01
16	2.000e+13	7.000e+01	3.000e+01	1.100e+03	1.213e+00

New data loa SPAYN 2.4.1.R Copyright © 2009 Simucad Design Automation, Inc.

Figure 14.7 Silvaco VWF data exported to Spayn. After Ref. [1].

14.1 Internet-Based TCAD Laboratory

Although the technology computer-aided design (TCAD) and electronic computer-aided design (ECAD) approaches to device design, technology development and very-large-scale integration (VLSI) circuit design are becoming popular, these are mainly restricted to R&D groups and semiconductor companies, and there is a barrier for the general CAD tool users since use of these CAD tools requires advanced knowledge of process, device physics and design. It is well known that laboratory experimentations are indispensable in engineering education, which however, traditionally been considered impractical for distance education. Hands-on experiments form a large portion of engineering education especially the undergraduate and postgraduate studies. Traditionally, these have been confined to hands-on proximal (face-to-face) experiments that require extensive resources including laboratory space and supervision. Creating new microelectronics curricula is a challenge. The creation of new curricula requires:

- Significant time and effort to develop a large number of needed subjects, which must include all the necessary materials—lectures, laboratories, subject projects, homework, examinations, etc.
- Expert knowledge of new design challenges and advanced integrated circuit (IC) design techniques.
- Experience in using best-in-class electronic design assistant (EDA) tools in the designs of contemporary ICs and their components.
- Development of educational design kits (EDKs) and process design kits (PDKs) which must correspond to real technologies to reflect their specific attributes, available with little to no restriction to universities.
- Comprehensible presentation to enable students to easily master the capabilities of contemporary methods of IC design and their solution methods.

A new conceptual framework for teaching technology and device design is necessary which will allow the student and teacher to interact more in the learning process by doing things in a new way—hands-on learning—by means of simulation. All aspects of technology cross sections can be illustrated by using the TCAD in process/device analysis and design. The importance is the fact that these simulations are immediately available to the teacher and student to rerun and hence explore the multidimensional space of physics and device design. Given a reasonable junior-level device/process course, this new methodology will provide the opportunity to initiate a strong interactive simulation environment for both experimenting with physics and technology. There is unique opportunity for students to embark on real device/technology design and independent study.

The Internet is an ideal medium for disseminating educational material to students and remote instruction purposes. With the globalization of education and the advancement of Internet technology, there has been a continuous move toward using more online resources in engineering education. The globalization of education has imposed extra pressure on courses with hands-on laboratory classes, which usually require the student's physical presence. Due to recent technological advances in computer technology and software, it is now feasible to implement more advanced and efficient, highly interactive, and very user-friendly systems without using very costly custom-written software and tools. One of the emerging uses of the Internet in engineering is to make the laboratory facilities available to the wider community.

Currents advancements in the computer and software technology are changing the roles of teachers and students. It is clear that the computer based educational technology has reached the point, specifically in the area of engineering education, where many major improvements have been made, such as significant cost reductions by possible replacement of the existing lecture-based courses, and the virtual instruments (VIs) may provide a highly interactive user interface and advanced analysis facilities that are not deliverable in the conventional methods. The challenges are:

- Providing engineering students with real world relevant laboratory experimentation
- How to actively engage students in nano- and microelectronic experimentation to gain a clear understanding

The literature in engineering education research suggests that a cohesive series of laboratory exercises improves the learning and retention of the material presented in a laboratory course. Currently, in most final-year undergraduate and virtually all master programs there are courses on device physics and processing technology (frequently as one single course) based on standard text books on metal–oxide–semiconductor (MOS) and bipolar device physics. There is a very strong connection to the device modeling area and typical process technology. For simulation, the students generally use device simulator (BIPOLE) or process simulator (SUPREM) or circuit simulator (SPICE) with probably no interactions between the process/technology and device parameters. In the attempt to cope with the increasing demand from distant students to have access to higher education in the field of engineering technology, universities worldwide have initiated activities to offer Internet-based laboratory educational programs. A great effort has been put in establishing a number of technologically and pedagogically advanced remote laboratory (RL) exercises. Some achievements in the development and use of such laboratories are presented below.

Recent advancements in the Internet, World Wide Web (WWW) technology, and computer-controlled instrumentation allow Internet-based techniques to be used for setting up RLs. Internet-based RLs are currently being used for experimental demonstrations to enhance classroom lectures, assignments, and homework in regular courses, and for laboratory experiments. For extending microelectronics education, attempts have been made to develop Internet-based device characterization laboratories.

Advantages of online laboratories are:

Currently many academic institutions offer a variety of web-based experimentation environments, the so-called RLs that support remotely controlled physical experiments. RL environment is a complete self-contained environment, which allows users to perform laboratory experiments for online distance learning programs. It is a creative step that bridges the gap between software simulation and the real physical laboratory experience. These new tools enable universities to provide students with free experimental resources without a substantial increase in laboratory cost. A review on RLs reveals the existence of RLs designed for teaching in different engineering fields throughout the world, being electronics, robotics, automation, physics, the fields that contain the largest amount of developments of RLs. In the majority of cases, the remotely accessible online laboratories reported in the literature are dispersed around the world and represent stand-alone systems.

The design of remote hardware-based laboratory experiments and their web implementation present different levels of complexity depending on the nature of the experiment and equipment necessary for running the experiment. For example, the development of a fully remote-controlled laboratory module (consisting of 15 experiments, say) experience goes well beyond the design and implementation of a single experiment. Classification and selection of experiments (e.g., a short- or a long-duration experiment) itself is a big issue in developing the experiment module. The design of experiments (DOE) must be pedagogically sound to ensure that students achieve the same level of performance (and experience) with respect to the learning outcomes as would be expected from a conventional laboratory. Thus the development of an RL module goes well beyond the design of the experimental apparatus and support software. RLs may be offered in three forms:

Hardware-based RLs provide students with web access to actual equipment. Physical RLs require space for equipment but less than hands-on laboratories, and they also use the real data acquired from equipment over the Internet.

Remote virtual laboratories (VLs) offer model simulation which needs minimum investment. Moreover, students can easily repeat experiments with the same process and results for analysis, although the experimental data may not be available.

Hybrid RLs combine both hardware-based experiments and remote simulation. Students can choose remote simulation or remote physical equipment-based experiments. This helps also possible comparison between the simulation results from the virtual simulation and the experimental data/results from physical equipment for the same experiment.

This new trend places demands for new design methodologies as well as new design tools. An advanced TCAD framework should have the following capabilities:

- Developing new process technology for device design and optimization

- Calibrating the process for a technology node so that the developed models can be used in the new process
- Providing process windows for given device performance targets through DOE for the final optimal design
- Providing a first-order approximation on device performance in relation to process variables
- Extracting circuit performance from virtual wafer experiments

In several chapters we have discussed on process and device simulation, hardware-based device characterization, and SPICE parameter extraction. Next, we describe an Internet-based integrated RL system which can be used for laboratory microelectronics education purposes. Briefly, virtual wafer fabrication (VWF) (described in the first part in this chapter) can now be realized with remote Internet-based hardware- and simulation-based laboratories. Here, we describe two laboratories:

- The *Microelectronics and VLSI Engineering Laboratory* (MVL) is a hardware-based remote microelectronics laboratory for device characterization.
- The *Technology CAD (TCAD) Laboratory* is a simulation-based laboratory dedicated for semiconductor device SPICE parameter extraction.

TCAD is a specialized field that deals with the design and simulation of microelectronic processes and devices. In the field of semiconductor TCAD, at present there is rarely any RL for educational purposes. It is also possible to integrate the above laboratory modules into a single online laboratory system so that the students can perform both the measurements and simulations, compare the experimental results with simulation and then extract SPICE parameters sequentially. The integrated online laboratory system will allow students to perform experiment at any time and from anywhere using a Java-enabled web browser through the Internet. As an educational tool, the laboratory platform will enable students and educators, who do not have access to conventional laboratories, to complement their theoretical knowledge by carrying out experiments and simulations remotely.

Microelectronics is a required core course in nearly every electrical and computer engineering curriculum. Electrical and computer engineers need a solid understanding of how to analyze and design circuits containing microelectronic devices (op amps, diodes and transistors). Although many of the concepts are not extremely difficult to visualize, the content is rich in design and associated trade-offs. Many of the trade-offs are complex due to the nonlinear nature of the active devices used. Integrating simulations and experiments is a powerful way to learn about a device. Students can compare measurements from real devices with predictions derived from the models they learn in class. By doing this, they gain a better understanding of the conditions under which these idealized models are applicable and where the models fail.

To meet the expanding needs for micro- and nanotechnology, new education programs and course modules must be developed and delivered to students at all levels. Due to the high cost of a microelectronic fabrication laboratory, teaching microelectronic circuit fabrication is very much driven by the availability of resources. Since 2009, efforts have been made to develop integrated microelectronics device characterization, parameter extraction and device/circuit simulation laboratory, a new online laboratory system that may significantly enhance the microelectronics laboratory education. MVL has been developed using high-end equipment. The simulation-based TCAD laboratory has been developed for the SPICE parameter extraction of semiconductor devices. The main focus of this work has been how to build low-cost RL experiments with the minimum hardware and software still maintaining satisfactory quality of the online experiments. Important issues such as, the development of an RL module consisting of several experiments, management of experiments, quality of service, security, safety and the operation of the equipment have been addressed in detail. In the first part, the focus has been placed on hardware-based RL experiments and their integration into a hybrid (both the measurement and simulation based) RL.

Different types of remote electronic laboratories are being made available via the Internet over the last few years. Design of online learning environment requires various elements to deliver instruction, facilitate interaction and to enhance the quality of learning. In the design phase one needs to decide on the target requirements and at the development of the technical and pedagogical solutions for the RLs. The fundamental objectives of engineering instructional laboratories are as follows:

- Appropriate instrumentation and/or software tools to make measurements of physical quantities.
- Appropriate experimental approach, specification of appropriate equipment and procedures, implementation of the procedures, and interpretation of the resulting data.
- Demonstration of appropriate levels of independent thought, creativity, and capability in real world problem solving.
- Identify issues related to equipment, experimental procedures and activities.
- Effective communication about laboratory work with teachers and students for comprehensive technical and laboratory reports.

Electronic/electrical measurement-based experiments may be conducted using different technologies. However, used technique could be unique to that experiment and also in the tools used to publish them over the Internet. Java is commonly used to provide a good base for Internet functionality and standardization across different computer hardware and operating systems (OSs). Choosing the best technology for remote-controlled laboratories development is dependent on several criteria such as, the current status of the technology, the development time, and the system independency of the user in terms of the OS and Internet browser as well as other necessary installations. Also the design and operation of the experiments are topic/domain dependent. Different tools are now available like, National Instruments (NI) LabVIEW (Laboratory Virtual Instrument Engineering Workbench) to publish the developed environment in the web. To minimize special purpose hardware and software investment, our system has been developed around LabVIEW software and NI hardware. LabVIEW program is one of the high level programming languages that is preferred in this study to perform VL applications. LabVIEW software provides an active and controllable interface for test, measurement, instrument control, data processing, and data analysis applications. Virtual instrumentation is a new technology paradigm which leverages all the continuing advancements in processor/PC technologies, bus standards, and Internet-based communication, software standards, protocols, and all application and network connectivity features. VIs are developed using LabVIEW to display process variables, set control parameters, and indicate the outcome. HP-VEE and other popular programming environments may also be used for VI development. Laboratory developers can either deploy the instrument through an online laboratory management system or develop their own management system using the package.

The first task after deciding on a computer one needs to do when setting up a cluster is to install an OS. Many considerations need to be made to arrive at an answer as to which OS is the best for the environment. Obviously, the most important consideration is whether or not the target application is supported under the OS in question. Sometimes the application is only supported under one OS and the choice is then unnecessary. When working with applications for the microelectronic industry, the choice will be between Microsoft Windows and at least one UNIX-based OS. Looking at the platform support from Silvaco is a fair example since it will be one of the software vendors utilized in the final design. There are numerous choices when it comes to purchasing hardware, which can be a daunting task for those who aren't familiar with the landscape. Hardware vendor flexibility should always be a major consideration in any of these decisions, especially in the storage arena.

To create a remote experiment, firstly we need implementation of local instrument control, and secondly we need implementation of remote instrument control through the Internet. To implement remote instrument control, interface architecture between the WWW server and LabVIEW is developed. Common Gateway Interface (CGI) and Transmission Control Protocol (TCP) are two major tools for communication between users and the WWW server, and the WWW server and the program running in other machines. Interface software has been developed for remote instrument control. After implementation of remote instrument control, one can set up the RL by designing the experiment content, such as user interface of the RL and instructions. The user follows the instructions to conduct the remote experiment. The system does not allow users to create the experiment by them and conduct it, the user can perform only available experiments. The major steps for remote online laboratory development involve:

1. Design the experiment.
2. Remotely control and operate the instruments.
3. Convert to web applications.
4. Launch the experiments on the Internet.

A modular online RL will typically consist of 10–12 hardware-based experiments which need to be made available always to students. As one particular experiment can be performed at a time by an individual student or a group of students (in case of collaborative learning) which require a careful scheduling of the experiments. Thus, for proper implementation and management of online laboratory system an efficient laboratory management system is essential. Administrative interface and database management should include the following:

Details of VL system operating specifications:

- Measurement, monitoring, and control interface
- Local operator interface and control of experimental setup
- Web server configuration
- System architecture for sharing RLs in campus and beyond
- DOE

Before the laboratory session:

- The experiment (timetable, students' list, equipment)

During the laboratory session:

- The experiment (operation and control of equipment)
- Batch experiments versus interactive experiments
- Time scheduling

After the laboratory session:

- The experiment (laboratory report preparation, submission)
- Student feedback
- Automation of student performance evaluation
- Evaluation and grading
- Laboratory grade sheet preparation/publishing

All movements (moving a physical equipment or making connections) in the conventional laboratory are done manually by the technical personnel. Some experiments in an electronics laboratory often require change in experimental setup which is accomplished in a conventional laboratory by the technical personnel. Robots can be programmed to perform tasks, such as, move forward and/or reverse. The development of robot assisted laboratories remotely accessible through the Internet is a rather complex task as it involves expertise from multiple disciplines such as remote operation of robotics, intelligent control, human–computer interaction and web-based intelligent programming.

14.2 Microelectronics and VLSI Engineering Laboratory Module

Measurements are fundamental to an understanding of any semiconductor device. Toward teaching device design and developing concepts, an online RL for characterizing microelectronic devices is described. Through a Java-enabled web browser, users from all over the world can run experiments on real transistors, diodes, and other devices by means of a semiconductor parameter analyzer Agilent 4145B or 4156C. The interface is simple and the result of the experiment is the voltage-current characteristic graph of the requested device. The tests are made in real time; the results shown are given as graphics and as tables and can be reviewed by the user at any time. Figure 14.8a shows the home page of the MVL module. The list of available experiments is also shown in Fig. 14.8b.

Optical microscopy and atomic force microscopy (AFM) are used extensively in physics, microelectronics, biotechnology, pharmaceutical research, mineralogy and microbiology. An optical microscope, often referred to as a light microscope, is a type of microscope which uses visible light and a system of lenses to magnify images of small samples. Though there has been significant progress in developing real-time microscopy and radiological image sharing over the Internet, this has not been extended to the domain of education. A remote optical microscope experimental setup is shown in Fig. 14.9.

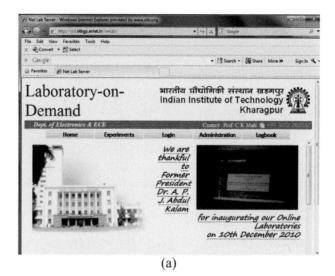

(a)

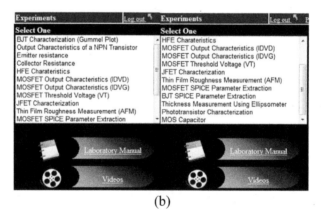

(b)

Figure 14.8 (a) Home page of the Microelectronics and VLSI Engineering Laboratory module. (b) List of available nano- and microelectronic experiments.

One can engineer an optical microscope which is operated via the Internet using most of the Internet and networking capable equipment, including laptops and desktop computers. The system offers a user the ability to conduct visual inspection of samples, microelectronic devices and manipulation experiments in the field of biology. The remote-controlled microscope is programmed to perform acquisition of images. The instrument is used to perform thin-film characterization experiments in a microelectronics laboratory. The server operates the microscope via the following steps:

1. Start the camera and obtain a picture.
2. Save the picture.
3. Prepare a histogram and send the data to the server.
4. Store the image in a folder on the web server.

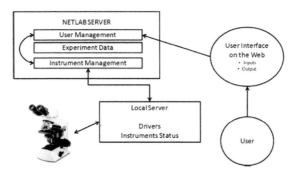

Figure 14.9 Remote optical microscope experiment setup: general configuration.

All the commands to operate and gather the data from the microscope is stored as a LabVIEW program. The program itself is hosted as a web services. This gives flexibility of access from different platforms. When the client calls a web service, the program is run and the data is returned. The histogram of the image is displayed along with the image is shown in Fig. 14.10. Apart from viewing the user can save and image on the server or download the image in the remote PC.

Design and implementation of a cost-effective atomic force microscope for remote online nanotechnology laboratory is described below. Based on a commercially available low-cost atomic force microscope (Model EasyScan2) and the NI LabVIEW graphical programming environment, we have developed the remote-controlled measurement system. The AFM characterization experiment is developed using NI LabVIEW which provides intuitive developing interface with the possibility of developing GUI applications. COM Automation interface of the EasyScan2 software has been used in our application. The EasyScan2 control software offers an interface where the user can automate a measurement task by writing a script which defines the task. The script objects specifically developed for the experiment access the online atomic force microscope which controls the scan range or feedback set point. The experiment takes a long time (typically 20 minutes) to complete the scan. It scans over a fixed area (1 μm × 1 μm) and the following information is collected:

- The image size
- The number of lines
- The number of points per lines
- The maximum amount of time in seconds to spend on a line
- The autoreset parameter in Boolean

All these inputs amount to a few bytes of data and can be sent as GET parameter to the web service.

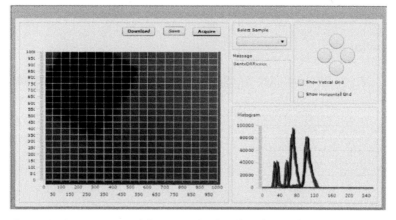

Figure 14.10   Histogram of the image displayed on the user's PC.

The experiment (scanned image of the surface of a thin film) yields data in several megabytes (see Fig. 14.11) and also it gets delayed during transfer to the user PC. Thus, the data has to be saved on the server PC itself for the user to fetch the data at a later time. The output of the experiment (an image) needs to be displayed as an image on the user PC. The image is reconstructed in 2D and enhanced by contrast stretching. It is kept in grayscale to minimize hard disk space consumption (see Fig. 14.12).

Figure 14.11 Scanned image of the surface of a thin film.

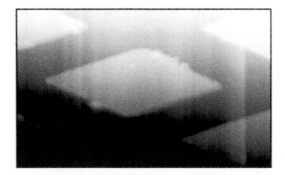

Figure 14.12 Scanned image of the surface of a thin film now reconstructed in 2D and enhanced by contrast stretching.

14.2.1 Integrated Technology CAD Laboratory

With the current maturity of TCAD tools, real wafer fabrication (RWF) can be emulated by process simulation from which realistic device structures can be generated; and the transistor performance can be characterized through device simulation with reasonable accuracy. At the heart of microelectronic engineering, one needs to be able to produce a device via some process, capture those steps into a concise computer model, and then run simulations to improve upon the design. VWF has become an integral part of semiconductor fabrication. For manpower training (both via lectures and simulation laboratory), several advanced educational institutions in the West have recently introduced courses on TCAD in their curricula. With the introduction of TCAD for microelectronic fabrication, it is imperative now that the teaching institutions in India introduce a new course/laboratory on TCAD for teaching [1].

Currently, no educational institution in India offers TCAD courses/laboratories due to nonavailability of such laboratories. Next, an integrated measurement-based MVL with a simulation-based TCAD laboratory is described. The main aim is to create infrastructure for a state-of-the-art TCAD laboratory online. As described in the MVL, the students perform real measurements on a wide variety of devices. The measured experimental data (device characteristics) can then be passed on to the simulation-based TCAD laboratory for the extraction of SPICE parameters. The home page of the integrated TCAD laboratory is shown in Fig. 14.13.

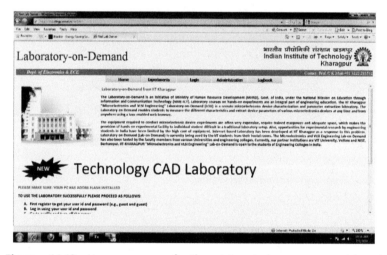

Figure 14.13 Home page of the integrated measurement-based Microelectronics and VLSI Engineering Laboratory with a simulation-based TCAD laboratory.

The transistor electrical characteristics are closely coupled with the doping profiles and layer structures, and very often, a trade-off must be taken for different design targets. The TCAD laboratory has been designed for developing two levels of knowledge: the definition/implementation of the simulation script (input file) and the analysis/understanding of the results. The current version of the laboratory module has the following sessions:

- Session 1: The process simulator SUPREM is introduced. The student simulate step-by-step a bipolar transistor. During this session, the doping profiles are investigated after each process step.
- Session 2: The device simulator BIPOLE is introduced. In this session the students compute the output characteristics of the bipolar transistor for several process parameters.
- Session 3: The student uses the SUPREM to design a process flow for a transistor. The students test the process flow by simulating it and extracting the doping profiles.
- Session 4: With the device simulator BIPOLE, the students compute the output and Gummel characteristics of the BJT for several process parameters. The electrical characteristics are used to obtain the compact model of the transistor. The student extracts Gummel-Poon model parameters from TCAD device simulations.

In this section, we describe the outcomes for some of the above tasks:

Process simulation is used during technology development to refine a process recipe, and during technology characterization to model the input structure for device simulation. Figure 14.14 shows the process simulation page using SUPREM. Modeling of processes provides a way for the student to interactively explore the fabrication process, study the effects of process choices, in a word, participate in the activity of a VFW "factory."

In the field of microelectronics, a device simulator is an important engineering tool with tremendous educational value. With a device simulator, a student can examine the characteristics of a microelectronic device described by a particular model. This makes it easier to develop an intuition for the general behavior of that device and examine the impact of particular device parameters on device characteristics. A device simulator lets students explore device behavior in regimes that would otherwise be infeasible or unsafe to examine. Device simulation is used to obtain simulated electrical measurements of a device, largely for technology characterization. Figure 14.15 shows the device simulation page for a bipolar transistor using BIPOLE device simulator.

Figure 14.14 Process simulation page.

Figure 14.15 Device simulation page.

Measurement is the final arbiter for any semiconductor simulation module. The developed TCAD laboratory includes several types of measurement systems as has been described in the hardware-based RL. Figure 14.16 shows the device characterization (output characteristics) page for a bipolar transistor measured using Agilent 4156C device parameter analyzer. The measured experimental data is then passed on to the simulation-based TCAD laboratory for extraction of SPICE parameters. Figure 14.16 also shows the extracted SPICE parameters.

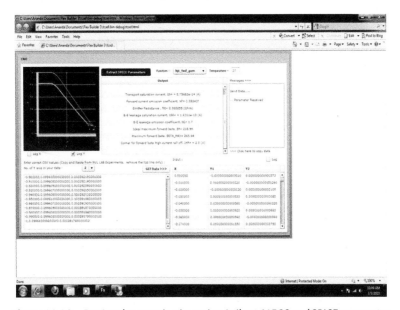

Figure 14.16 Device characterization using Agilent 4156C and SPICE parameter extraction.

14.3 SPICE Parameter Extraction

The highest level in the proposed TCAD laboratory module is the SPICE parameter extraction. The main link between circuit-level simulation and lower-level TCAD tasks is through the compact models used in circuit simulators to characterize the behavior of individual circuit components. Simulators have tremendous educational value. With a device simulator, a student can examine how the device described by a particular model behaves when presented with various inputs. Although this exploration can be done with a real device, the appropriate equipment is often prohibitively expensive. These models are fitted to the data produced from process and device simulation, providing circuit designers with a CAD environment that accurately characterizes the manufacturing lines that will make the circuits. In the TCAD laboratory, the exercises are designed to expose the students to experimentation with real devices and extraction of SPICE parameters. Ordinarily, students are unable to practice these skills in conventional device simulation laboratory. In the online TCAD laboratory described above, a simulation program, Web-based Tool for Electronic Model Automation (WEBTEMA), is used for the extraction of SPICE parameters (WEBTEMA, 2010). The software is an EDA tool for automated SPICE modeling of advanced semiconductor devices.

The complete set of semiconductor dc parameters can be quickly and accurately evaluated with Agilent 4156C stand-alone semiconductor device parameter analyzer. As an example, the measured Gummel plot of an NPN transistor is shown in Fig. 14.17. The measured data can be graphically analyzed to obtain saturation current, current gain, and current gain versus collector current characteristics, along with base resistance and recombination current characteristics. The extracted SPICE parameters from the measured Gummel plot are also shown in the Fig. 14.17.

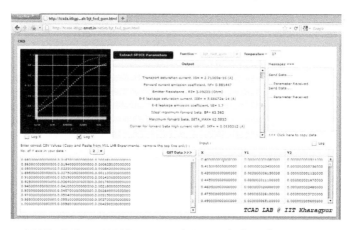

Figure 14.17 Agilent 4156C–measured Gummel plot of an NPN transistor along with the extracted SPICE parameters using the WEBTEMA tool.

14.4 Summary

A systematic study based on TCAD is taken up for the design and VWF of metal–oxide–semiconductor field-effect transistors (MOSFETs) in complementary metal–oxide–semiconductor (CMOS) technology. The impact of different types of variability that affect CMOS technology has been addressed. The impact of the virtual wafer technology on new process development has been described. Supplementing expensive iterative experimentation of VLSI technology development by VWF and virtual device characterization is demonstrated. A full-factorial DOE (TCAD simulations) are executed and the process compact model (PCM) is extracted. From calibrated TCAD, it is possible to generate a response surface model and find the optimized 45 nm CMOS process. The process is optimized (to select a stable process window) with respect to threshold voltage, drive current, and leakage. Successful implementation of VWF technology could lead to the general services to the chip design and fabrication industry with the idea of Internet-based TCAD. It represents a novel idea of the virtual fab foundry, which is a first attempt at bridging the gap between chip design and chip fabrication, and it has potential commercial applications in ECAD based on TCAD, as well as professional services.

The online microelectronics laboratory for professional scientists and engineers in advanced device and process technologies and, in particular, for the engineering curriculum has been discussed. We have described the design, implementation, and functionality available on hardware-based RL platforms in order to run a number of experiments. The design and implementation of a remotely controlled online (Internet-based) experiment on thin-film characterization for a nanotechnology laboratory using an optical microscope and an atomic force microscope have been described. The prototype online AFM characterization and optical microscopy experiments developed showed the feasibility of such systems for a low-cost shared-resource remote nanoelectronics laboratory for education. This work shows that use of a process simulator to teach semiconductor processing is indeed possible via an online TCAD laboratory. The proposed new methodology will provide an opportunity to initiate a strong interactive simulation environment for experimenting with both physics and technology. There is a unique opportunity for students to embark on real device/technology design via independent study.

The proposed RL can be used not only in the field of microelectronics education but also for doing any measurement-based task with real laboratory instruments. Furthermore, RLs are not intended to replace available hands-on experiments but rather to give students the possibility to acquire their own data sets for further analysis and to think on the way how device/technology design should be done.

Reference

1. Gruener, C. J. (2009). *Design and Implementation of a Computational Cluster for High Performance Design and Modeling of Integrated Circuits*, MS thesis, Rochester Institute of Technology.

Index